INDIANA

Advancing and Driving the Economy

The Pyramids, part of the 200-acre College Park office and commercial development in northwest Indianapolis.

INDIANA
Advancing and Driving the Economy

Steve Kaelble

This book was produced in cooperation with the Indiana Manufacturers Association. Cherbo Publishing Group gratefully acknowledges their important contribution to *Indiana: Advancing and Driving the Economy.*

cherbo publishing group, inc.

president	JACK C. CHERBO
chief operating officer	ELAINE HOFFMAN
editorial director	CHRISTINA M. BEAUSANG
managing feature editor	MARGARET L. MARTIN
senior feature editor	TINA G. RUBIN
senior profiles editor	J. KELLEY YOUNGER
profiles editors	NEVAIR KABAKIAN
	LIZA YETENEKIAN SMITH
associate editors	SYLVIA EMRICH-TOMA
	JENNY KORNFELD
editorial assistant/proofreader	MARK K. NISHIMURA
profiles writers	SYLVIA EMRICH-TOMA
	NEVAIR KABAKIAN
	JENNY KORNFELD
creative director	PERI A. HOLGUIN
senior designer	THEODORE E. YEAGER
designer	NELSON CAMPOS
senior photo editor	WALTER MLADINA
photo editor	KAREN MAZE
digital color specialist	ART VASQUEZ
sales administrator	JOAN K. BAKER
client services supervisor	PATRICIA DE LEONARD
senior client services coordinator	LESLIE E. SHAW
client services coordinator	KENYA HICKS
executive assistant	JUDY ROBITSCHEK
administrative assistant	BILL WAY
regional development manager	GLEN EDWARDS
eastern regional manager	MARCIA WEISS
publisher's representatives	TIM HOOVER
	JEANNE SCHEDEL

Cherbo Publishing Group, Inc.
Encino, California 91316
© 2008 by Cherbo Publishing Group, Inc.
All rights reserved. Published 2009.

Printed in Canada
By Friesens

Subsidiary Production Office
Santa Rosa, CA, USA
888.340.6049

Library of Congress Cataloging-in-Publication data
Kaelble, Steve.
A pictorial guide highlighting Indiana's economic and social advantages.

Library of Congress Control Number 2008943261
ISBN 978-1-882933-12-9
Visit the CPG Web site at www.cherbopub.com.

The information in this publication is the most recent available and has been carefully researched to ensure accuracy. Cherbo Publishing Group, Inc. cannot and does not guarantee either the correctness of all information furnished it or the complete absence of errors, including omissions.

To purchase additional copies of this book, contact Joan Baker at Cherbo Publishing Group: jbaker@cherbopub.com or phone 818.783.0040 ext. 27.

Acknowledgments

This book is dedicated to my family, the center of my life in Indiana. My thanks go out to all who have welcomed me to the Hoosier State through the years and taught me so much about its wonderful history, amazing business accomplishments, and incredible people. I'm grateful to the CEOs who have invited me into their offices and given me insights into their business success and the civic leaders who have shared with me the stories of their outstanding communities. And finally, special thanks to the Cherbo Publishing Group team, including my editor, Tina Rubin, who ensured that the stories I'm telling reflect both the state's fascinating heritage and its bright future.

—Steve Kaelble

The seven-story Indianapolis Artsgarden exhibit and performance space over the intersection of Washington and Illinois streets.

TABLE OF CONTENTS

PART ONE — 2
TAKING SUCCESS IN STRIDE:
AN OVERVIEW OF INDIANA

CHAPTER ONE
INDIANA LIVING: Diverse and Affordable — 4

CHAPTER TWO
FABRICATED, FASHIONED, FORGED:
Manufacturing — 22

CHAPTER THREE
A PLACE TO GROW:
Agriculture, Forestry, and Food Processing — 34

CHAPTER FOUR
BUILDING TOMORROW'S KNOWLEDGE BASE:
Higher Education — 44

CHAPTER FIVE
VITAL SIGNS:
Life Sciences and Health Care — 54

CHAPTER SIX
LINKED TO THE WORLD:
Logistics, Transportation, and Energy — 64

CHAPTER SEVEN
BUILDING BLOCKS OF SUCCESS:
Real Estate, Construction, Finance, and Insurance — 74

PART TWO — 84
SUCCESS STORIES:
PROFILES OF COMPANIES AND ORGANIZATIONS

PHOTO CREDITS — 136

715 Locust Street

Civic Centre Auditorium & Convention Centre
Auditorium
715 Locust Street

From left to right: Chase Tower in Indianapolis; Evansville Auditorium and Convention Centre; the 1931 Coca-Cola Bottling Company in Indianapolis.

From left to right: An Indiana Pacers NBA game against the New York Knicks at Conseco Fieldhouse; Lucas Oil Stadium, home of the NFL's Indianapolis Colts and the 2012 Super Bowl; exuberant Fighting Irish after a touchdown against the Boston College Eagles at Notre Dame Stadium in South Bend.

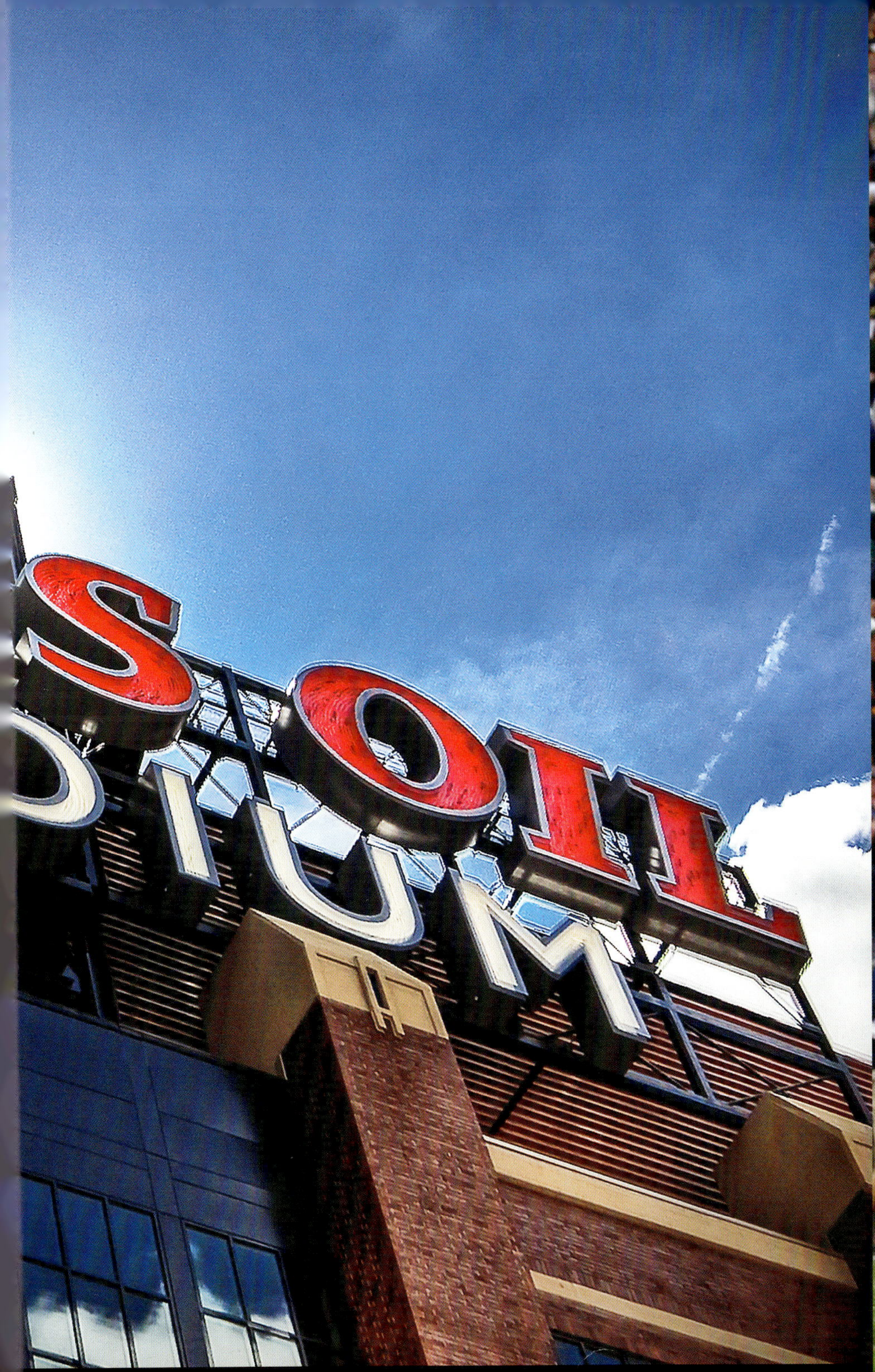

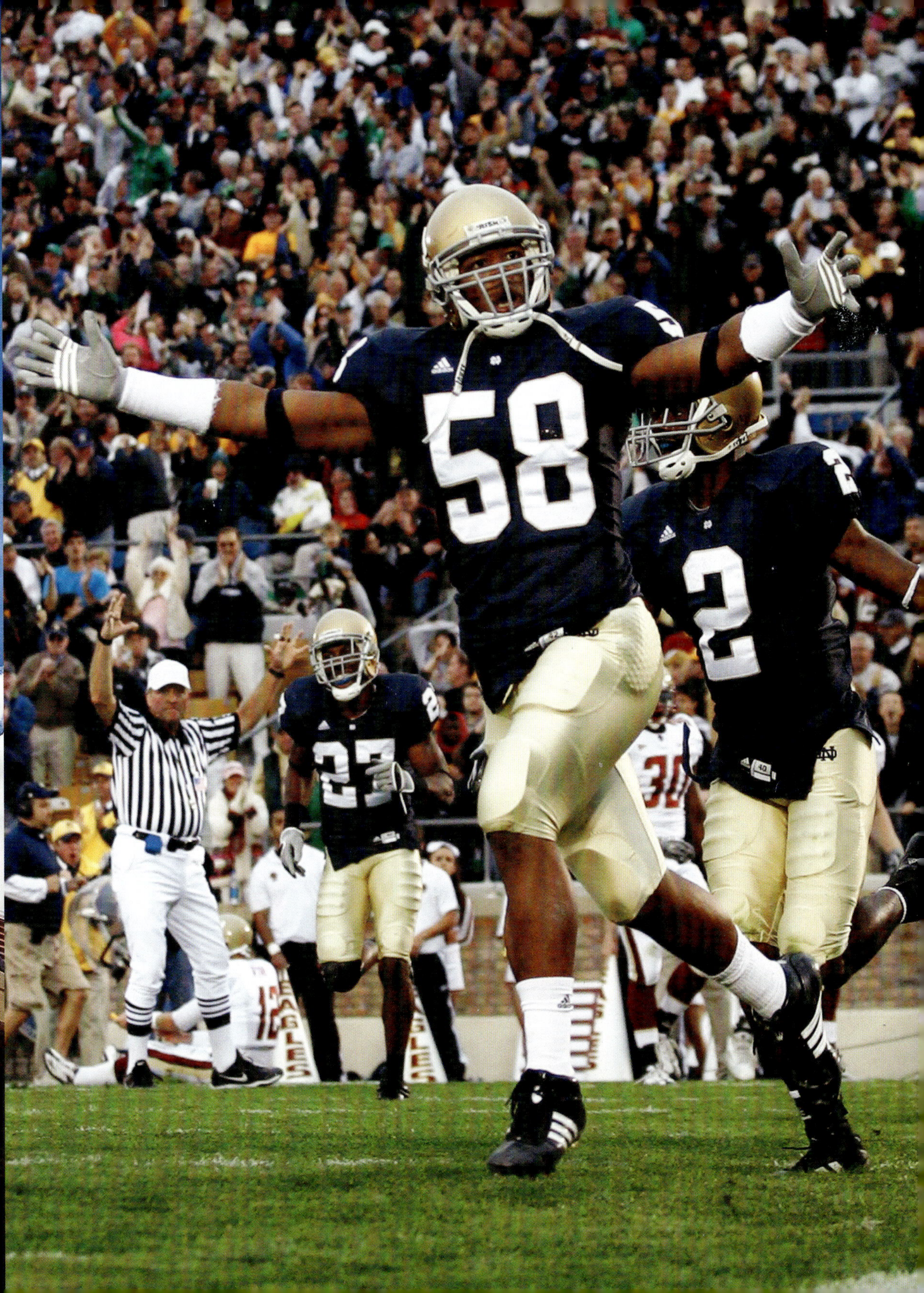

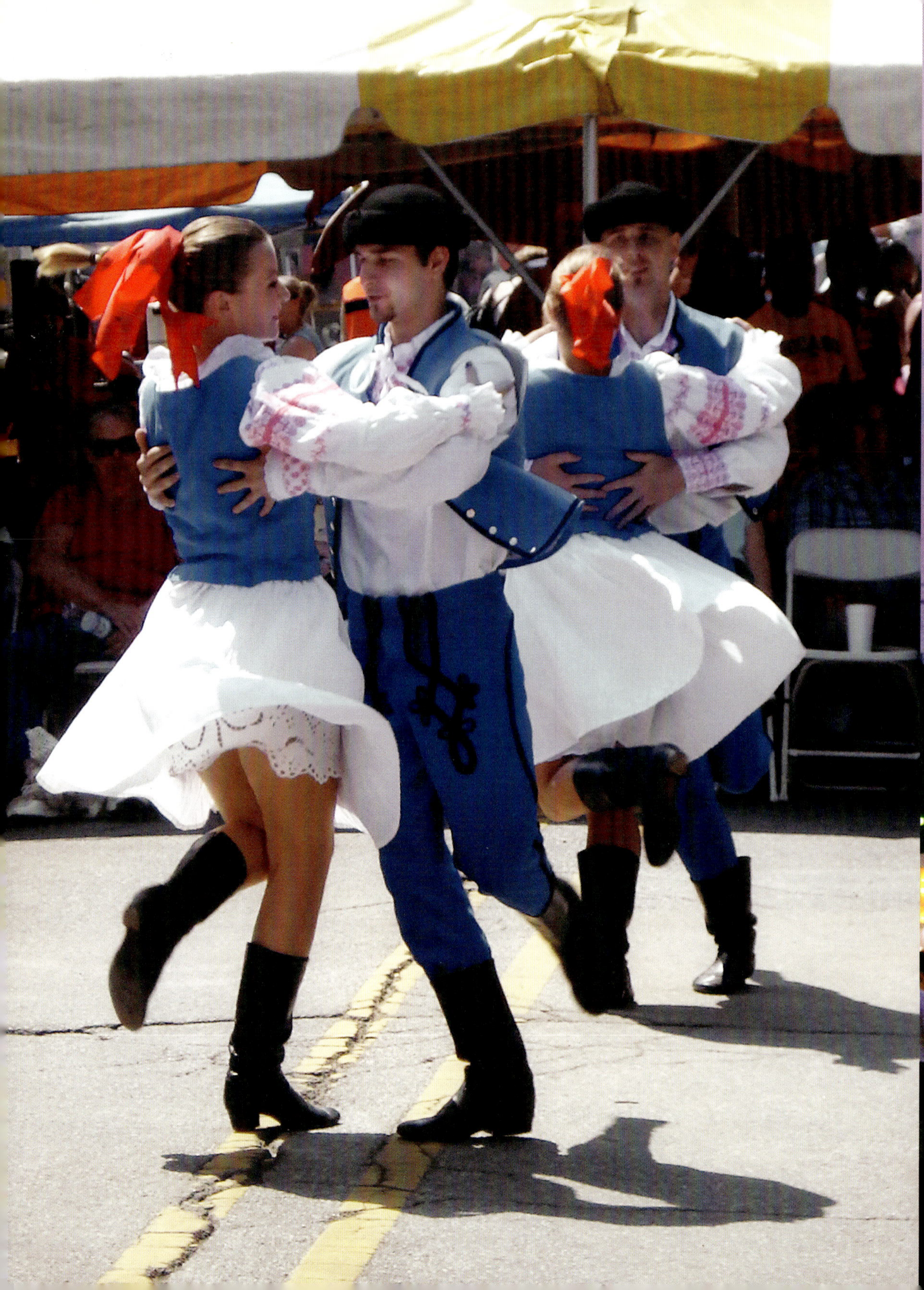

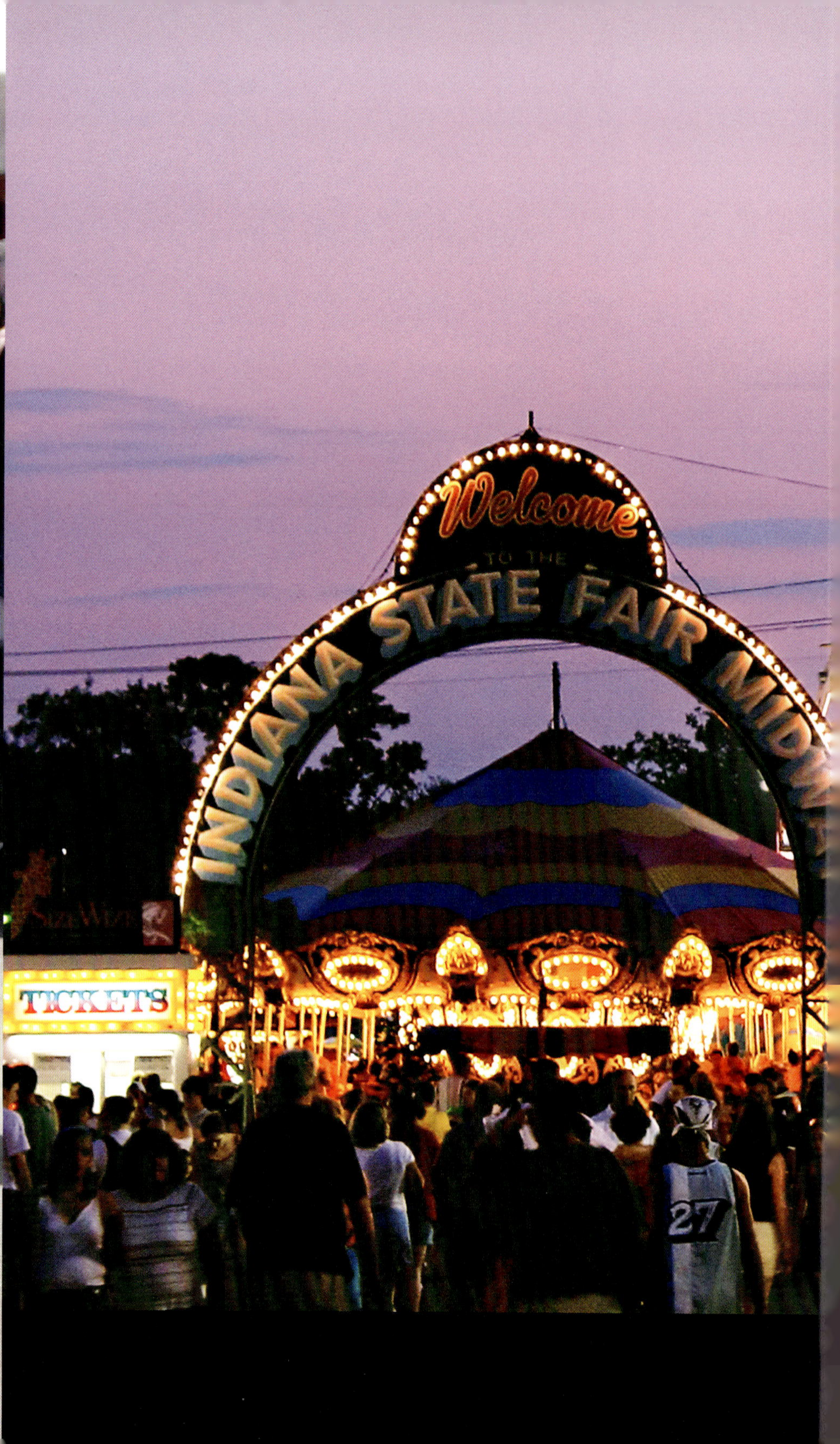
Welcome
TO THE
INDIANA STATE FAIR
TICKETS

From left to right: Polish dancing at Pierogi Fest 2008 in Whiting; the evening crowd at the Indiana State Fair in Indianapolis; trying a hand at pottery during the Indiana Festival at the Conner Prairie Living History Museum in Fishers.

xiii

From left to right: The Nate Sutton Quintet (shown: Nate Sutton and Courtney Crouse) performing at Tutto Bene Wine Cafe & More in Bloomington; admiring the art at a benefit auction in Bloomington; *Wisdom Keepers*, a bronze by Bruce LaFountain at Eiteljorg Museum of American Indians and Western Art in Indianapolis.

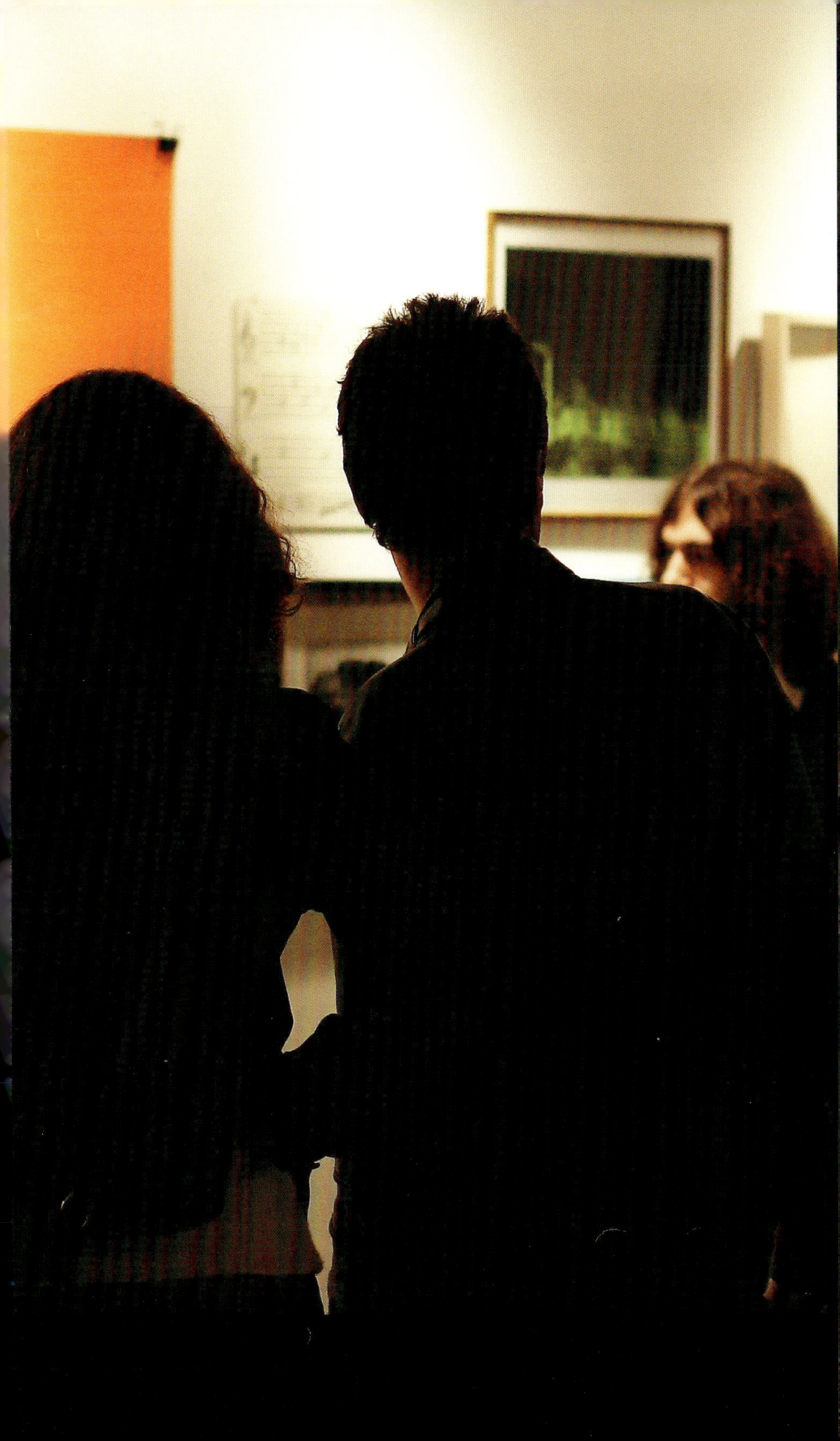

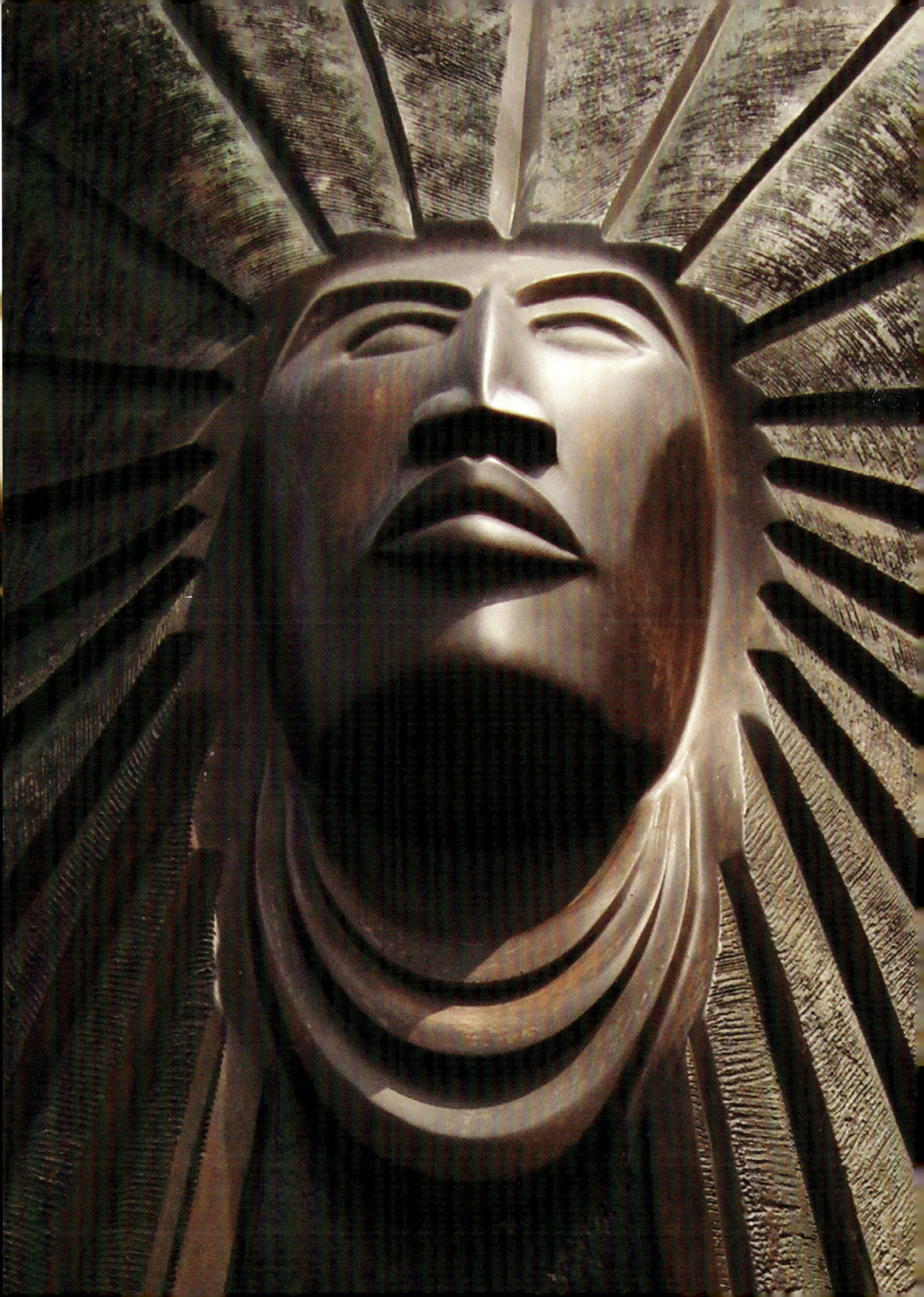

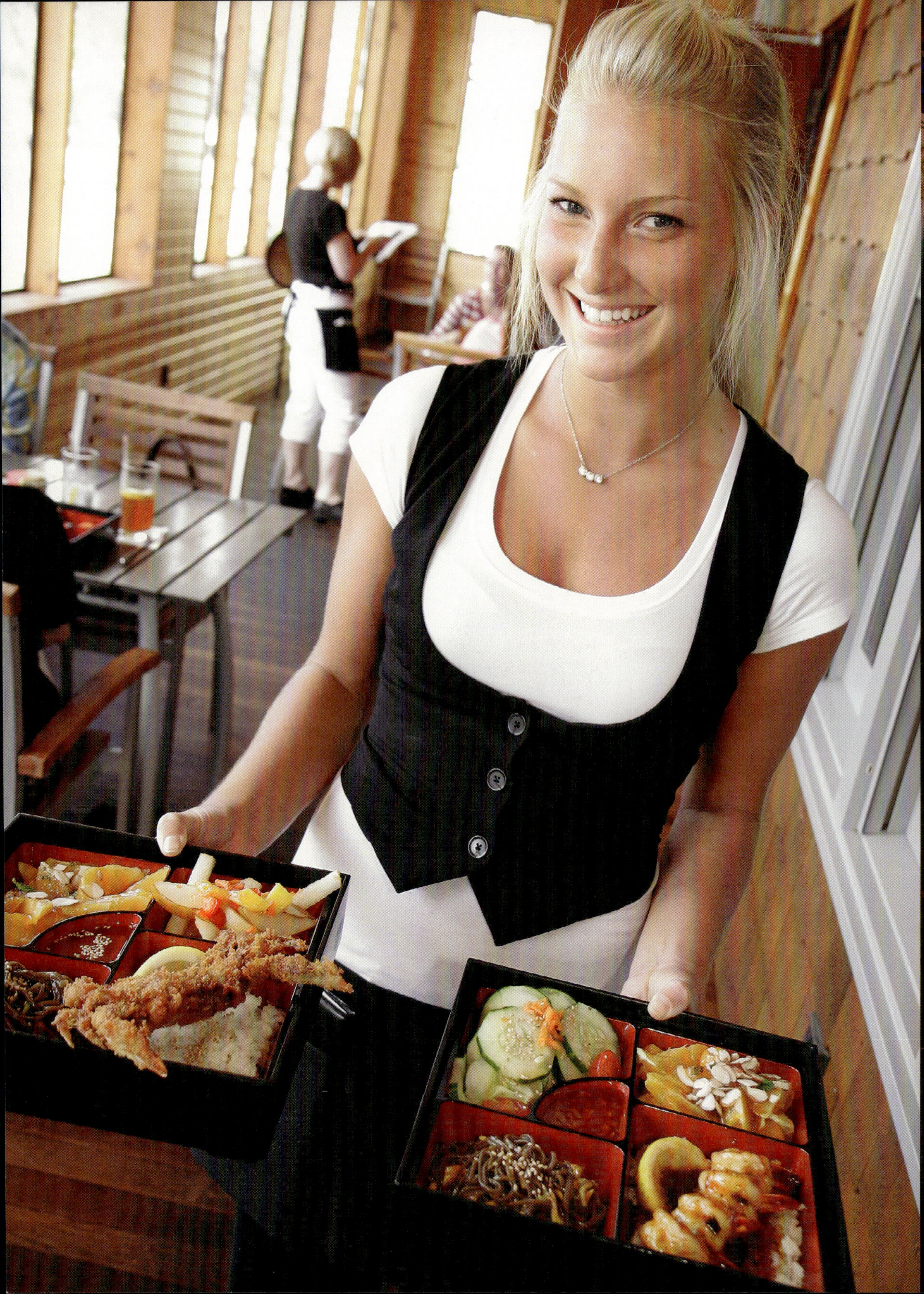
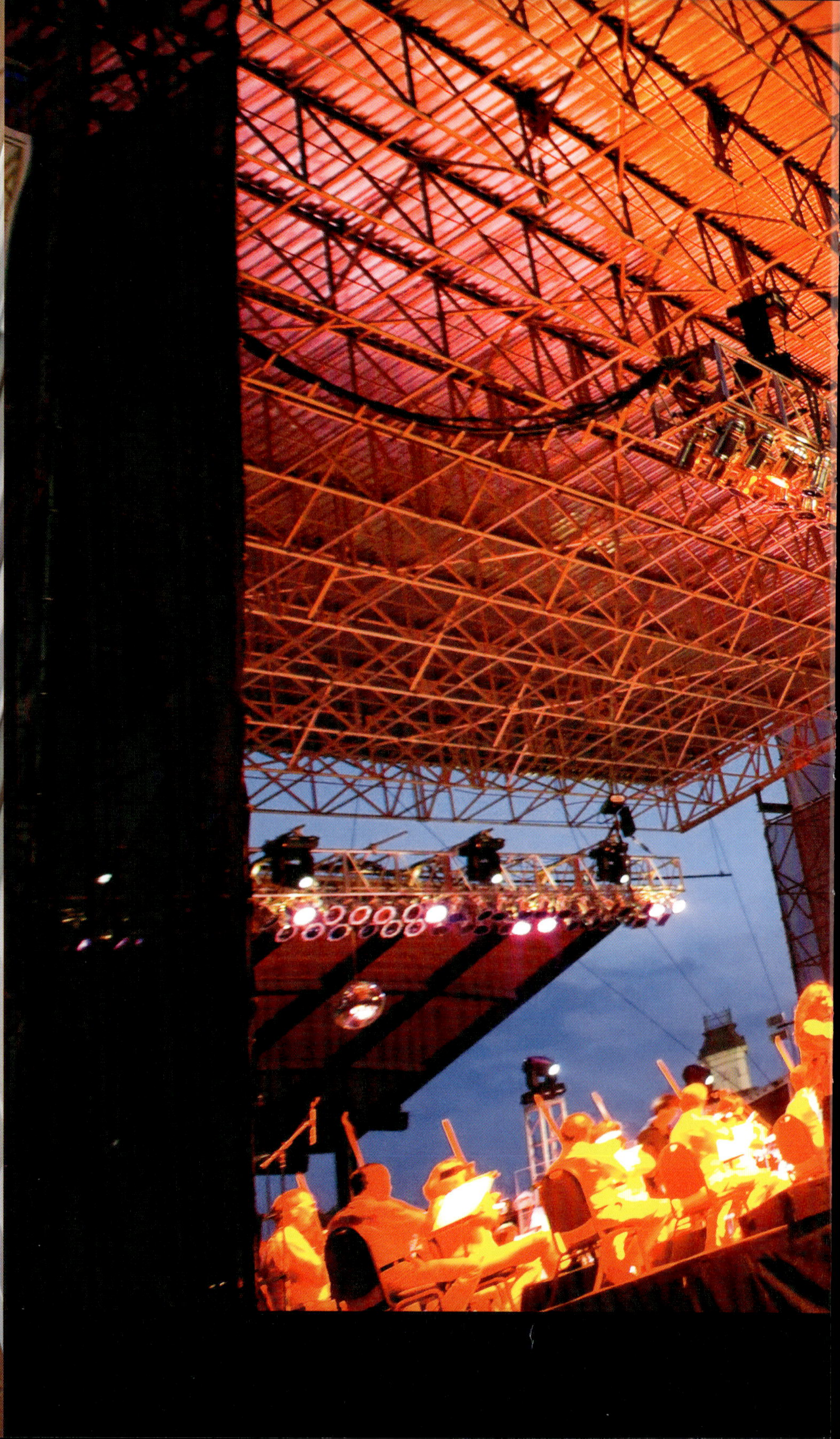

Opposite page, left: An inspired presentation of Asian fusion food at the Cerulean Restaurant in Winona Lake, outside Fort Wayne. This page: An evening concert

From left to right: A Shipshewana farmer loading bales of hay from wagon to (unseen) truck; Indiana grain traveling by barge on the Ohio River between

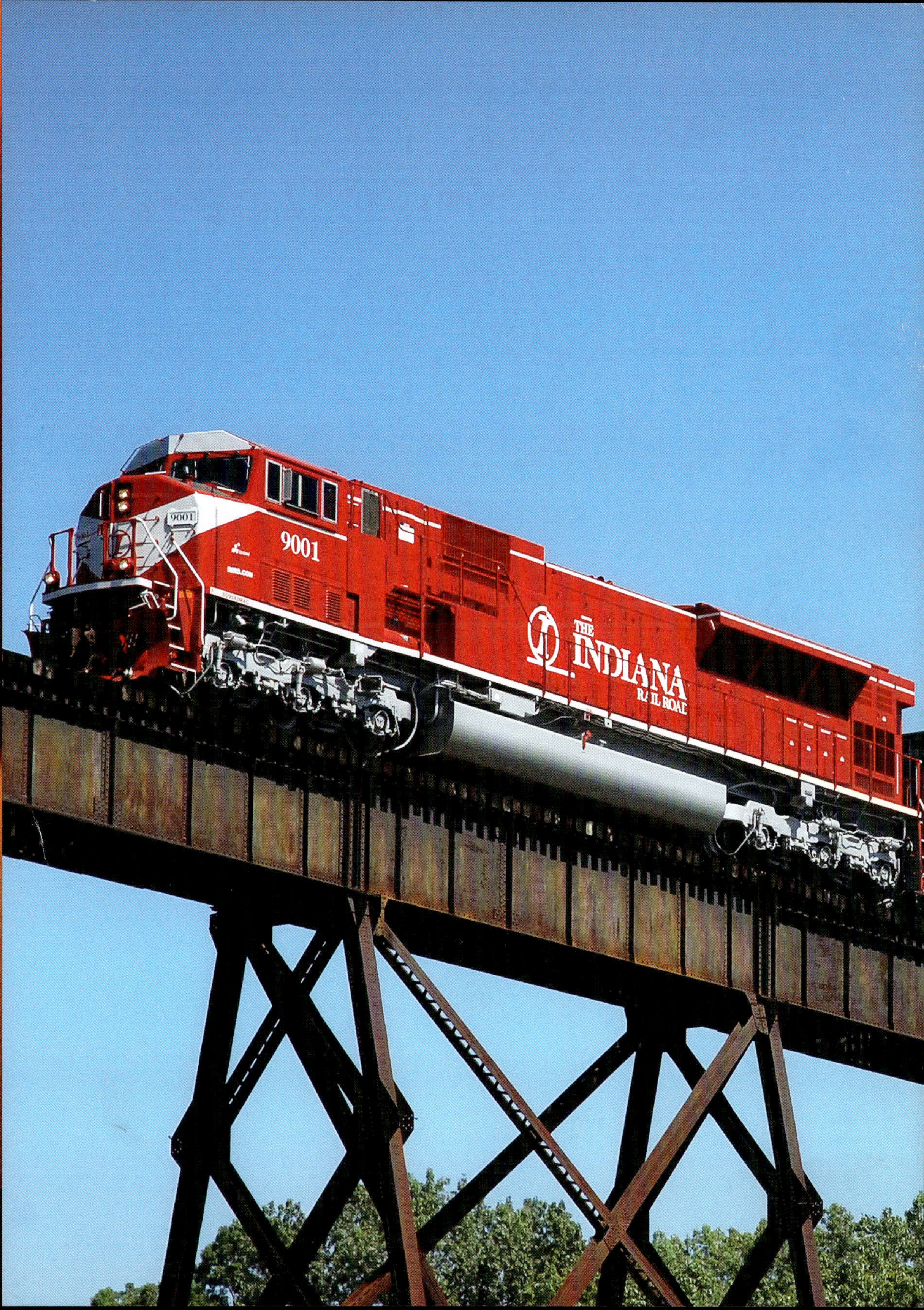
9001
THE
INDIANA
RAIL ROAD

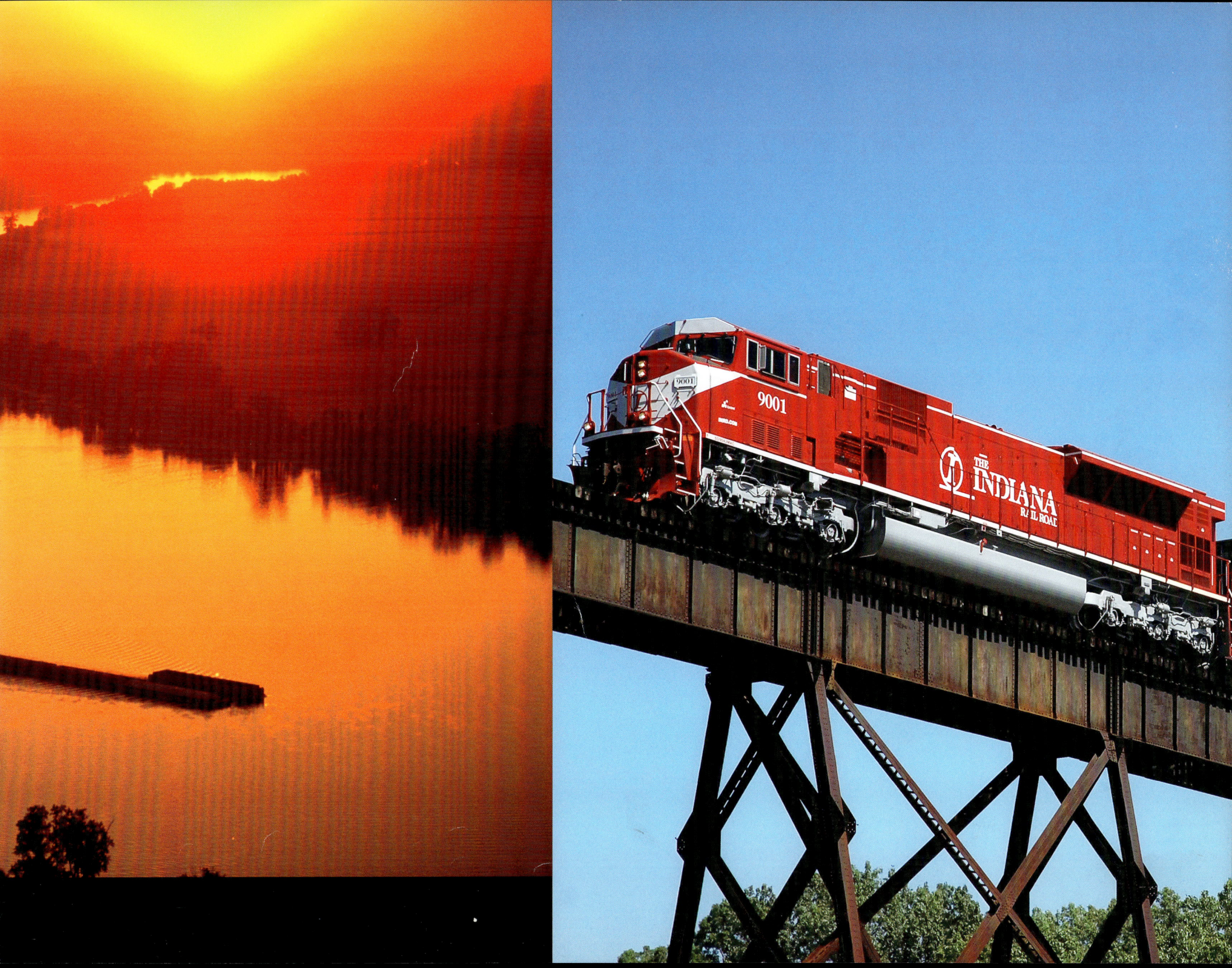

CORPORATIONS & ORGANIZATIONS PROFILED
The following organizations have made a valuable commitment to the quality of this publication.

BP Products North America, Inc. . . . . .xxi, 96–97

Courtyard by Marriott Indianapolis—Carmel, The 134

Duke Energy Corporation 100

Eli Lilly and Company 124–25

French Lick Resort . . . . .xxi, 130–31

Hyatt Place Indianapolis Airport 132

Hyatt Place Indianapolis Keystone 133

Indiana American Water 106–07

Kirby Risk Corporation 118–19

MasterBrand Cabinets, Inc. 116–17

Northern Indiana Public Service Company (NIPSCO)—A Division
   of NiSource Inc. 102

Old Hickory Furniture Company 112–13

ProLiance Energy 98–99

Roche Diagnostics 126

Sony DADC . . . . xxi, 88–89

Steel Dynamics, Inc. 110–11

United States Steel Corporation . . . . .xxi, 114–15

Vertellus Specialties Inc. 120

Vincennes University . . . . .xxi, 92–93

BUSINESS VISIONARIES

The following companies and organizations are recognized as innovators in their fields and have played a prominent role in this publication, as they have in the region.

BP Products North America, Inc.
Whiting Refinery
2815 Indianapolis Boulevard, Whiting, IN 46394
Web site: www.bp.com

French Lick Resort
8670 West State Road 56, French Lick, IN 47432
Phone: 812-936-9300
Web site: www.frenchlick.com

Sony DADC
1800 North Fruitridge Avenue, Terre Haute, IN 47804
Contact: Shelley Klingerman, Manager, Marketing & Communications
Phone: 812-462-8260
E-mail: shelley.klingerman@sonydadc.com
Web site: www.sonydadc.com
"One customer driven supply chain delivering global excellence through our people."

United States Steel Corporation
Gary Works
One North Broadway, Gary, IN 46402-3101
Phone: 219-888-2000
Web site: www.ussteel.com
"Making Steel. World Competitive. Building Value."

Vincennes University
1002 North First Street, Vincennes, IN 47591
Phone: 800-742-9198
Web site: www.vinu.edu
"Indiana's First College"

The new passenger terminal at Indianapolis International Airport, opened in November 2008.

FOREWORD

The Indiana Manufacturers Association welcomes you to *Indiana: Advancing and Driving the Economy*. As you will see, Indiana has vibrant, varied, and dynamic business and academic communities that create a quality of life enjoyed by over six million Hoosiers.

Here in Indiana, we know that business is the lifeblood of any economy. That is why we make it easier for businesses to succeed, with lower business costs, shovel-ready sites, a skilled workforce, and industry initiatives designed to stimulate growth, innovation, and profits.

We know that businesses want to keep their costs low. That is why Indiana has a tax structure that is extremely competitive. Our costs for industrial electricity are the second lowest in the nation. On top of that, Indiana has great economic incentives to help your business grow.

Indiana is called the "Crossroads of America" for good reason. We have more intersecting interstates than any other state, three international airports, a robust rail system, and three marine ports on Lake Michigan and the Ohio River.

Indiana is the most manufacturing-intensive state in America. Indiana is a leader in steel, automotive, pharmaceuticals, electronic and medical devices, furniture, recreational vehicles, and so much more.

The Indiana Manufacturers Association was formed in 1901 as an advocate for Indiana's business community. We are proud of the economy that makes up our Hoosier State. That is why we are pleased to showcase this excitement in *Indiana: Advancing and Driving the Economy*. We encourage you to come and enjoy all that Indiana has to offer.

IMA

Indiana Manufacturers Association

A couple planning their future as they stroll happily through autumn leaves in Terre Haute's Deming Park.

TAKING SUCCESS IN STRIDE:
AN OVERVIEW OF INDIANA

PART ONE

This page: The downtown Indianapolis skyline at twilight, seen from a footbridge above Canal Walk. Opposite page: Kayaking on the man-made East Race Waterway through downtown South Bend.

CHAPTER ONE

Quality of Life

INDIANA LIVING: DIVERSE AND AFFORDABLE

The diversity that is life in Indiana is revealed in many ways. Consider the modern St. Vincent Health headquarters building in Carmel, created by Indianapolis native Michael Graves as an urban structure with exterior patterns that reflect the state's agricultural landscape as it appears from the air. Contrast Indiana homes, ranging from comfortable, small-town bungalows and chic downtown condos to elite lakefront mansions. Share a scenic two-lane road with a horse-drawn Amish buggy or board a jet at the brand-new Indianapolis International Airport. Indiana's diverse surroundings serve as an apt metaphor for the widely varying styles of life that Hoosiers enjoy.

Indiana calls itself the Crossroads of America, for reasons relating both to location and infrastructure. The state is situated relatively near the center of the American population—in fact, the population center *was* in Indiana from the 1890s–1940s, and even with western population growth still remains not far away in eastern Missouri. That means it's comparatively easy for Indiana residents and businesses to reach Americans elsewhere, and doing so is made all the easier by the presence of solid infrastructure, from numerous interstate highways and modern airports to well-connected rail lines and ports that service huge, oceangoing container vessels—which one might not expect to find serving a landlocked state.

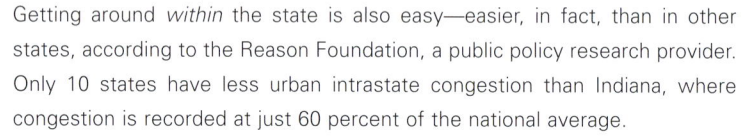

Getting around *within* the state is also easy—easier, in fact, than in other states, according to the Reason Foundation, a public policy research provider. Only 10 states have less urban intrastate congestion than Indiana, where congestion is recorded at just 60 percent of the national average.

The state's central location and business-friendly climate allow companies large and small to thrive. Indiana is home to 14 Fortune 1000 corporations, including WellPoint, Eli Lilly and Company, Cummins, NiSource, and Conseco. Indiana has given birth to fast-growing companies honored by *Inc.* magazine, from Endangered Species Chocolate in Indianapolis to recreational vehicle (RV) component maker Genesis Products in Elkhart. Businesses are unveiling innovations in the steelmaking process, creating new ways to collect Internet information, and bringing new life sciences products to market.

Indiana companies trying to attract talent from the coasts find they have a relatively easy sell, thanks to dramatic cost-of-living differences. For example, a San Francisco resident earning $150,000 a year would have to make just $84,450 in Indianapolis to maintain the same standard of living. For the first quarter of 2008, the ACCRA cost of living index for Indiana was 93.8, below the national benchmark of 100 and far below the 139.4 percent index in California and the 130.6 index in New York.

If the American Dream is owning a home, then the dream is alive and well in Indiana. According to the U.S. Census Bureau, the home ownership rate in the state was 73.8 percent in 2007, well ahead of the national average of 68.1 percent. And while 26.4 percent of all housing units nationwide were in multiunit buildings in 2000 (the latest data available), the percent in Indiana was only 19.2. The Hoosier State has traditionally fared well on housing affordability rankings, and the nature of its real estate scene is comparatively conservative. Home prices tend to appreciate slowly but surely, avoiding the biggest booms but also steering clear of the biggest busts. Hoosier homeowners weren't spared the impact of the residential real estate implosion of 2007–08, but the negative effects were milder than on the coasts.

Indiana's attractive attributes caught the attention of *Forbes* magazine when its editors sought to list America's best places to raise a family in 2008. Hamilton County, located just north of Indianapolis, ranked first on the list. Editors culled the possibilities by looking for the best schools, then checking the cost of living, home prices, property taxes, crime rates, and commute times. Hamilton County scored well in nearly all categories, yet has an average home price of $200,000, far less than many other attractive places.

Dynamic Settings

Indiana was named for the Indians who worked the land centuries before the United States was even a dream: first the mound builders, who left their mark at Angel Mounds near Evansville and Mounds State Park in Anderson, then a variety of tribes who left their names on places around the state, such as Miami, Delaware, Potawatomi. Settlers arrived as the American frontier opened, and because most came by boat, the earliest population centers were in the southern part of the state near the Ohio River. Indiana became the 19th state in 1816, with Corydon as its capital. By 1821 enough people had settled to the north that lawmakers decided a more centrally located capital would be appropriate, and a site at the confluence of White River and Fall Creek was chosen as the location of the new capital, to be called Indianapolis.

Opposite page, left: A historic home in Crawfordsville, northwest of Indianapolis. Opposite page, right: Sisters enjoying a spring day from their special vantage point. This page: Mounds State Park in Anderson.

The state's pioneer spirit lives on at such sites as Conner Prairie, a living history museum in the suburban Indianapolis town of Fishers that re-creates early Indiana life. But where it may be felt the most is in the rural work ethic, a legacy of the earliest settlers and farmers. To this day, it serves manufacturers and other employers far from the cornfields.

Many of Indiana's 6.3 million people live in large cities, but many more are scattered widely, in slower-paced small towns and rural areas. The biggest population center is Indianapolis, with about 785,600 residents within city limits and a total of 1.5 million in the metropolitan area. Fort Wayne is home to 249,000 people, while 116,000 live in Evansville, 105,000 in South Bend, and 98,000 in Gary. The fastest-growing town is Fishers, which recorded 38,000 residents in 2000 and 66,000 in 2007.

But the map is dotted with more modest cities, communities, and towns, including Laconia, which has a population of 29, and New Amsterdam, listed by U.S. Census estimators as having just one resident. The atmosphere of Indiana is as heavily influenced by quaint courthouse squares as it is by lively urban centers. What many find particularly appealing is that in Indiana they can enjoy the best of both worlds, by living in such small communities as Noblesville or Newburgh and being only a short drive from a busy downtown.

Seven distinct regions characterize the Indiana landscape, comprising wooded hills, farmland, urban areas, and lovely lakefronts.

This page, top: A Civil War reenactment at Conner Prairie living history museum in Fishers. This page, bottom: Up close and personal with a walrus at the Indianapolis Zoo. Opposite page: The Carmel Arts and Design District.

First, consider central Indiana. It may sound rural, but it actually encompasses one of the region's hottest places to live, which is decidedly urban: Indianapolis. The downtown area has evolved in the past two decades from a quiet, nine-to-five workplace to a center of work, play, and living that bustles nearly 24 hours a day. Housing ranges from trendy lofts and condominiums carved out of historic buildings to modern high-rises with upscale, penthouse living, and more units are being added every day. The area is also home to high-rise office buildings, busy shops (anchored by Circle Centre, a thriving mall located near the Indiana Convention Center), and enjoyable places within White River State Park such as the Indianapolis Zoo, White River Gardens, and the NCAA Hall of Champions.

Suburban living in central Indiana can include urban-style amenities, a village atmosphere, or anything in between. Affluent Carmel, for example, created the Carmel Arts and Design District, a collection of art galleries, specialty shops, and restaurants, with a state-of-the-art performing arts center under construction and slated for completion in 2010. The city is known across central Indiana for its widespread use of traffic roundabouts to ease congestion, even on some of the biggest thoroughfares. To the west of Carmel, the village of Zionsville is a quieter enclave, marked by shops along the bricked main street, which the city has meticulously maintained. Small-town flavor surrounds Indianapolis proper in such county-seat communities as Greenfield, in Hancock County, and Franklin, in Johnson County.

The glacier that flattened northern Indiana eons ago stopped its trek just south of Indianapolis, leaving south-central Indiana with lovely rolling hills that remain, for the most part, heavily wooded. This region, an appealing blend of culture and beauty, includes Bloomington—home to Indiana University, characterized by limestone buildings created from stone quarried nearby. The lively lifestyle that comes from the presence of a Big Ten university includes local and touring musicians, diverse sports and recreational activities, and colorful political activism. The region also includes Brown County and its central town of Nashville, a destination known for spectacular fall foliage, live entertainment, and down-home shops that showcase local arts and crafts. Columbus, the seat of Bartholomew County, has benefited tremendously from its generous corporate community, including the Fortune 500 company Cummins. A forward-thinking civic program launched in the early 1940s provided money for public entities and churches to hire world-class architects to design their new buildings. Today Columbus is home to more than five dozen buildings by prominent architects and is sometimes referred to as a museum of modern architecture.

The presence of the slow-moving Ohio River influences the character of the southern region of the state. The waters flow past the stately riverside mansions of communities rich in history. Madison, for example, prospered in the 19th century through steamboat and river trade; today more than 130 of its blocks are listed on the National Register of Historic Places. The communities of New Albany and Jeffersonville are part of the Louisville, Kentucky, metro area and have access to the city and all its amenities by way of Ohio River bridges. To the southwest is Evansville, which anchors the regional economy with its manufacturing and financial services industries. Two institutions of higher learning, the University of Southern Indiana and University of Evansville, impart a collegiate ambience to the city.

The western region brims with colleges and universities that draw students from around the world, and residents enjoy the quality of life that comes from having such schools nearby. The commercial hub of the western edge of Indiana is Terre Haute. While workers there are manufacturing high-tech digital media, plastics, and baking powder, students are being educated at the area's four colleges: Indiana State University, Rose-Hulman Institute of Technology, Saint Mary-of-the-Woods College, and Ivy Tech Community College–Wabash Valley. In the towns of Greencastle and Crawfordsville, rival schools DePauw University and Wabash College, respectively, educate the future workforce. And in Vincennes, students prepare for an associate's degree at Vincennes University, which in 2006 added the 800-seat Red Skelton Performing Arts Center, named after the internationally known Hoosier comedian whose hometown was Vincennes. The biggest concentration of students in this region, however, is in the Lafayette–West Lafayette area, home to Purdue University. The university influences local life in many ways. Faculty families encourage excellence in the local primary and secondary schools, university research sparks new jobs, and international students and faculty bring enriched cultural offerings.

Opposite page: Fall foliage in south-central Indiana's Brown County. This page: The Commons Mall, designed by architect Cesar Pelli, in Columbus.

Sometimes called The Region and known as a part of Chicagoland, the communities of northwestern Indiana, though they have strong economies of their own, serve as lower-cost bedroom communities for Chicago. It's less than 20 miles by highway or commuter rail from the Indiana community of Hammond, for example, to U.S. Cellular Field, home of the Chicago White Sox, and just a few miles farther to the Sears Tower. Residents of the area also enjoy the Indiana Dunes National Lakeshore along Lake Michigan, with its sand dunes up to 180 feet high, several large public marinas, and beachfront housing. The area boasts numerous colleges, too, including Valparaiso University, Calumet College of St. Joseph in Whiting, St. Joseph's College in Rensselaer, three regional campuses of Indiana and Purdue universities, and four Ivy Tech Community College campuses.

In northern Indiana, South Bend, on the St. Joseph River, is known worldwide as the home of the University of Notre Dame, attracting thousands of football fans throughout the fall. But the city also serves as a commercial and cultural hub of the region, with arts, sports, entertainment, and thriving businesses. Its attractive downtown includes one of the more unusual water features to be found in an urban setting: the East Race Waterway, a man-made whitewater rafting course. Fort Wayne, the state's second-largest city, is situated near the confluence of three rivers and brings a solid manufacturing base, growing technology sector, and high quality health care to northern Indiana residents. A slower pace characterizes the region's Elkhart County, a major Amish settlement affording a laid-back atmosphere and the opportunity to buy handcrafted merchandise reflecting old-world tradition.

This page: The Fort Wayne skyline at dusk, including the historic Allen County Courthouse. Opposite page, left: Kicking up a storm at Indiana Dunes National Lakeshore along Lake Michigan. Opposite page, right: The University of Notre Dame marching band at a home game in South Bend.

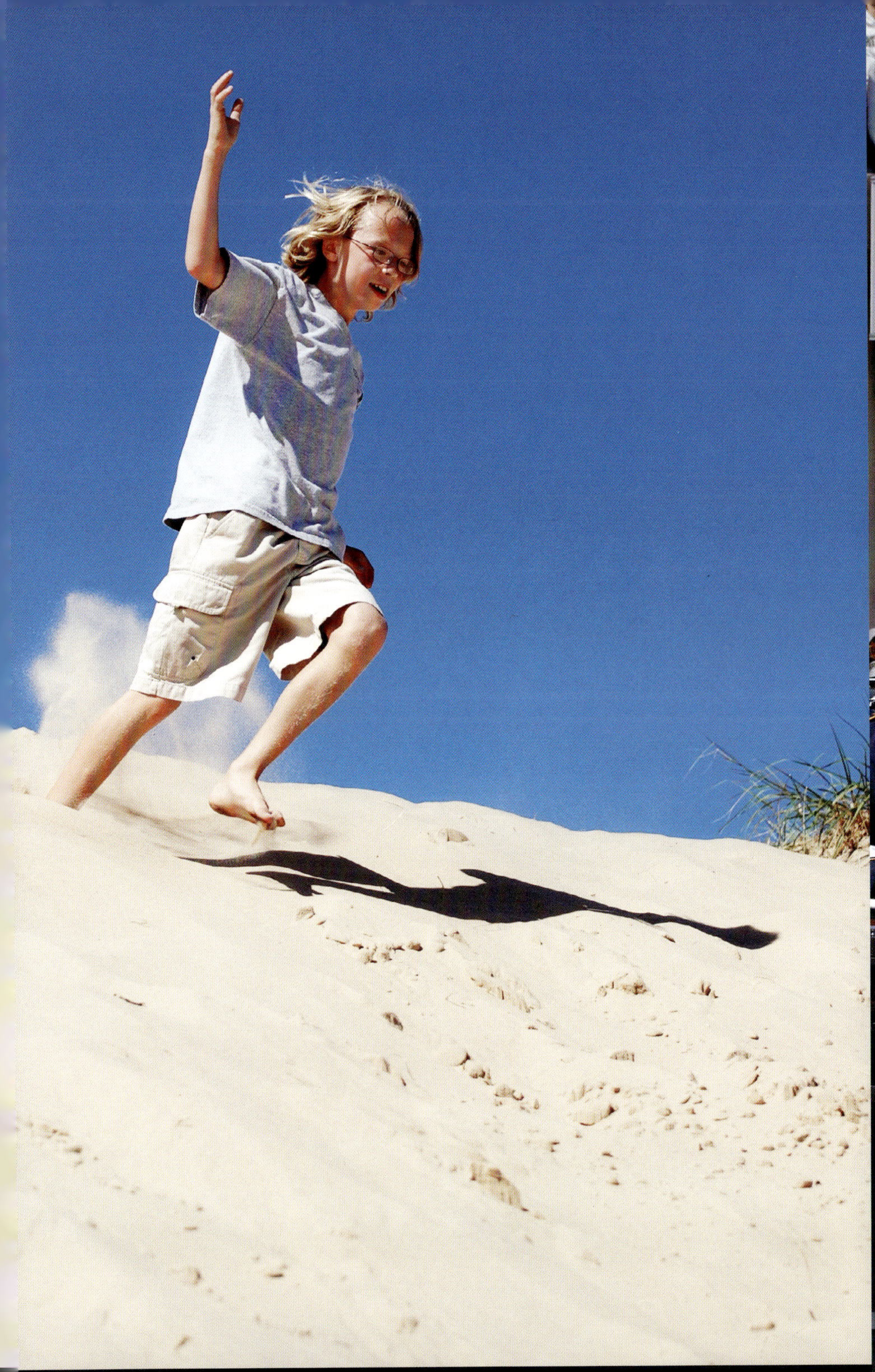

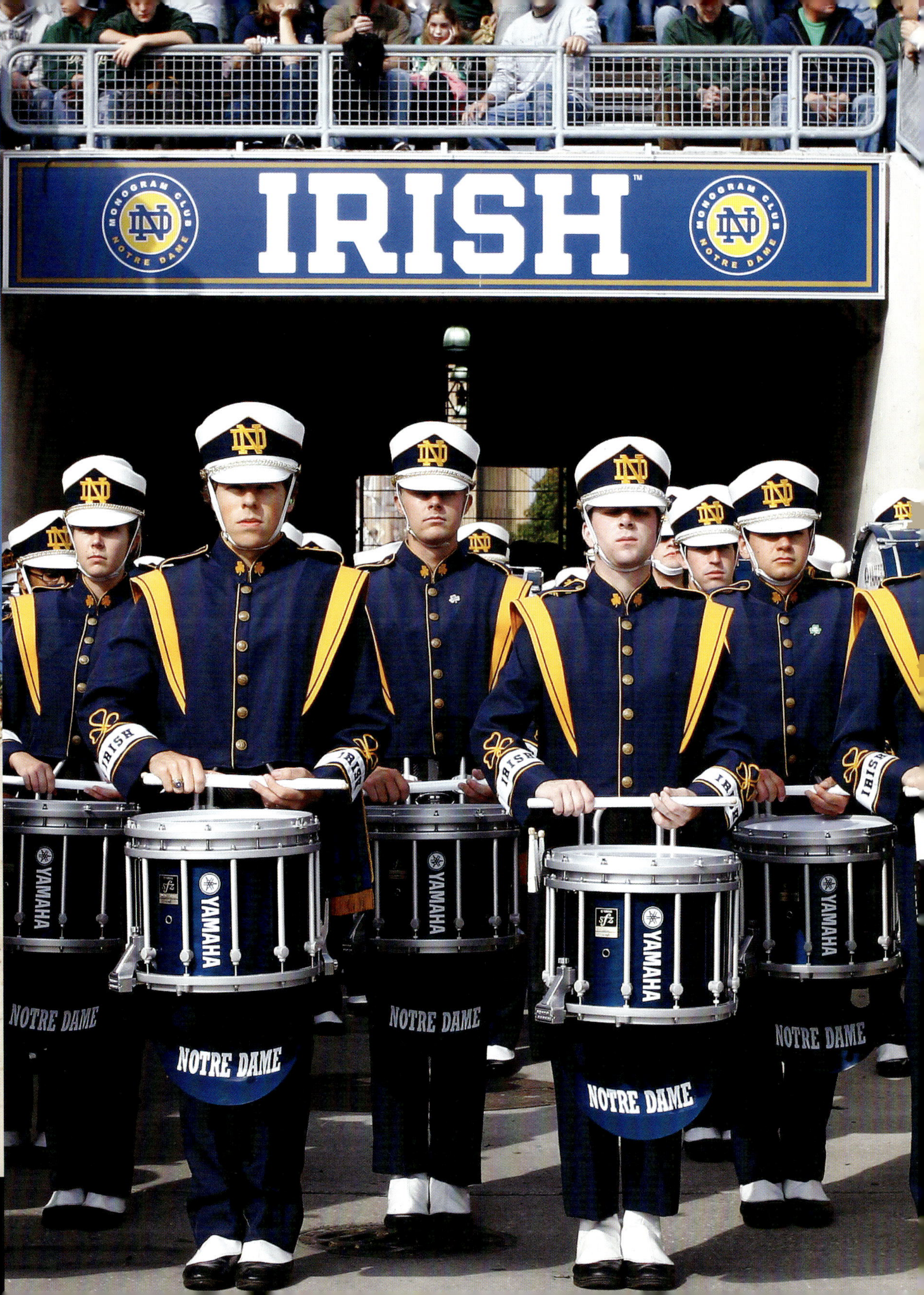
IRISH
MONOGRAM CLUB
NOTRE DAME
YAMAHA
NOTRE DAME

Small-town charm and college-town amenities can be found in eastern Indiana. Richmond's heritage even includes jazz, thanks to the appearance more than a century ago of Starr Piano Company and then Gennett Records, which served such prominent stars as Jelly Roll Morton, Hoagy Carmichael, and Bix Beiderbecke. College students from near and far study in Richmond, at Earlham College and local branches of Indiana University and Ivy Tech State College. The region is also home to 18,000-student Ball State University, consistently named to the *Princeton Review*'s list of the Midwest's best colleges, in Muncie. In addition to the university's economic impact, manufacturing gives Muncie a boost. Adding a further dash of fame is Garfield the Cat, a comic-strip creation born in Muncie three decades ago who has gone on to film and television success. A significant cottage industry revolves around the lazy comic feline, positively impacting the local economy.

Sports and the Great Outdoors

What better feather can a sports-crazy city put in its cap than one emblazoned with the words "Super Bowl Champions"? How about one that says "Welcome to the Super Bowl"? After the Indianapolis Colts won professional sports' most hyped event in 2007, their home city landed the rights to host the 2012 Super Bowl, in the new Lucas Oil Stadium. Other major league sports teams calling Indiana home are the Indiana Pacers of the National Basketball Association and the Indiana Fever of the Women's National Basketball Association. Sports fans also turn out by the thousands to cheer on Minor League Baseball's Indianapolis Indians, Fort Wayne Wizards, and South Bend Silverhawks, plus independent pro ball teams Gary SouthShore RailCats and Evansville Otters. The Indiana Ice of the U.S. Hockey League bring Tier One hockey to the Pepsi Coliseum at the Indiana State Fairgrounds.

The upcoming 2012 Super Bowl is just the latest in a long string of prominent sporting events scheduled in Indianapolis. Lucas Oil Stadium will host the National Collegiate Athletic Association's Men's Final Four in 2010, the sixth time one of college sports' biggest events has been held in Indianapolis since 1980 (more than in any other city). Indianapolis frequently hosts major amateur sporting events, including the 2008 U.S. Olympic Team diving trials. And if there's one thing for which the Hoosier State is known around the world, it's motor sports. Some of racing's biggest events take place in Indianapolis, including the National Hot Rod Association's Mac Tools U.S. Nationals, at O'Reilly Raceway Park on Labor Day weekend, as well as three huge events at the Indianapolis Motor Speedway: NASCAR's Allstate 400 at the Brickyard in July, the brand-new Red Bull Indianapolis Grand Prix motorcycle race in September, and of course, the world-famous Indianapolis 500 on Memorial Day weekend.

Opposite page, left: A statue of composer Hoagy Carmichael on the Gennett Walk of Fame in Richmond. Opposite page, right: Ball State University in Muncie. This page: A pit stop during the 2008 NASCAR Allstate 400 at the Brickyard, Indianapolis Motor Speedway.

This page: Canoeing toward the channel between Upper and Lower Fish Lakes in LaPorte County. Opposite page, left: Posing with angels' trumpets at the Foellinger-Freimann Botanical Conservatory in Fort Wayne. Opposite page, right: Fort Wayne Golf Association Men's City Championship at Brookwood Golf Club.

On the college level, Indiana is home to one of the most loyalty-inspiring football programs in the country, that of the University of Notre Dame, as well as a pair of Big Ten programs at Indiana and Purdue universities. What surprises some outsiders is the level of enthusiasm that also greets high school sports. Some of Indiana's high-school basketball arenas are as big as major college facilities, including one that seats 10,000 in New Castle and another that seats 9,000 in Anderson.

Natural beauty draws Hoosiers outdoors, particularly to the 24 state parks and national lakeshore. Among the popular state parks are Brown County in Nashville, Indiana Dunes on Lake Michigan, Clifty Falls near Madison, Spring Mill near Mitchell, and Fort Harrison in Indianapolis. All offer hiking and many have campgrounds and lodges. Pokagon State Park in Angola even has a refrigerated toboggan run.

Lakes and reservoirs, from Lake Michigan to such man-made reservoirs as Monroe, Brookville, Geist, and Patoka Lake, also serve as recreational magnets, and summer homes surround scores of them across the northern portion of the state. Natural beauty can be found inside city limits as well, at such major botanical gardens as White River Gardens in Indianapolis and Foellinger-Freimann Botanical Conservatory in Fort Wayne.

Golfers can find challenges in all corners of the state, with many courses designed by the game's top architects. In fact, renowned course architect Pete Dye hails from Indiana and owns a home on the 18th hole of one of his masterpieces, Crooked Stick Golf Club in Carmel. Because of his presence, Indiana has many outstanding Dye creations. The state also has a pair of courses crafted by Jack Nicklaus, Sycamore Hills Golf Club in Fort Wayne and the Sagamore Club in Noblesville, and an outstanding Robert Trent Jones–designed course, Otter Creek Golf Club, in Columbus.

Off the soccer fields and baseball diamonds, children have plenty of ways to keep busy and entertained outdoors. Indiana is home to a variety of zoos, including the Indianapolis Zoo, Fort Wayne Children's Zoo, Evansville's Mesker Park Zoo & Botanic Garden, and Michigan City's Washington Park Zoo, as well as two amusement parks: Holiday World & Splashin' Safari, in Santa Claus, and Indiana Beach Amusement Resort, in Monticello.

Celebrating Culture

The state's rich sports heritage may be well known outside Indiana, but its arts scene, perhaps less so. Indiana has a remarkably diverse collection of cultural offerings, from outstanding museums and performing arts organizations to eclectic festivals.

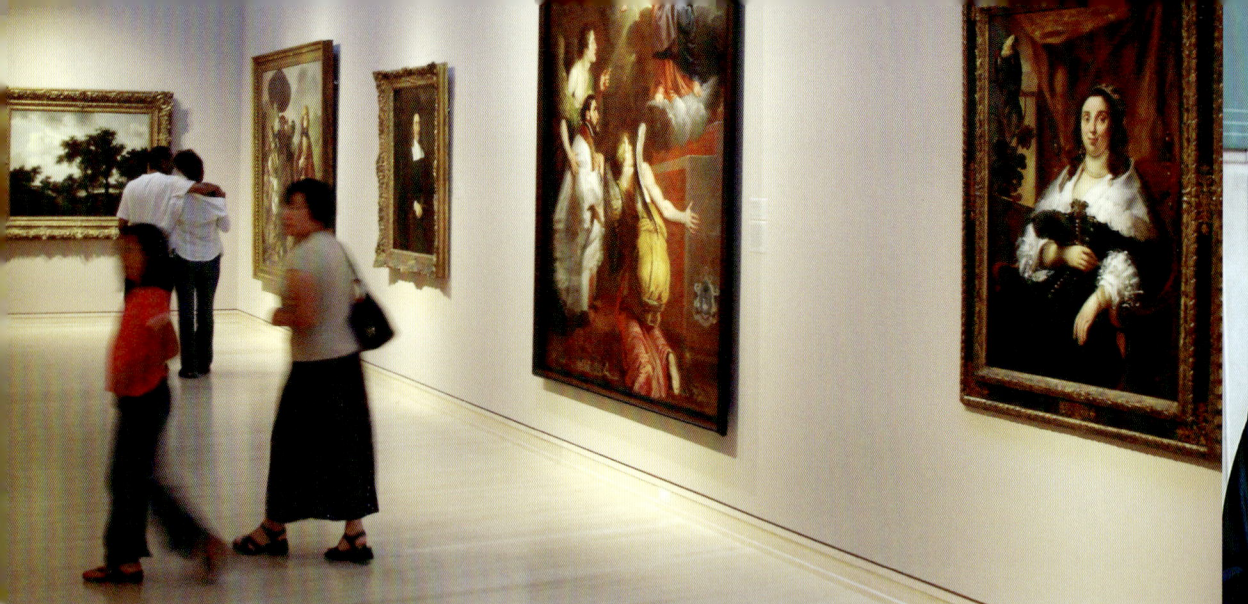

The Indianapolis Museum of Art is a world-class facility situated on a stunning piece of property once held by the Lilly family of pharmaceutical fortune. Its collections range from Asian, African, European, and American art to textiles and decorative art from a variety of periods. The museum's meticulously landscaped grounds, ideal for strolls, marriage proposals, or weddings, feature the Oldfields-Lilly House and Gardens. One hundred acres of meadows and wetlands adjacent to the museum campus are being groomed to be part of the Virginia B. Fairbanks Art & Nature Park, slated to open in 2009. Eight site-specific artworks, the first of many, have been commissioned for the park.

The Eiteljorg Museum of American Indians and Western Art, part of White River State Park in downtown Indianapolis, has an impressive collection of Western and Native American art, both contemporary and traditional, and the architecture of the building itself makes it appear as though it were transported from the Southwest.

There are other significant galleries, too. The Fort Wayne Museum of Art showcases American works created after 1850. The Sheldon Swope Art Museum in Terre Haute emphasizes Wabash Valley artists of all eras and 20th-century American painting and sculpture, with works by Edward Hopper, Grant Wood, and Thomas Hart Benton. The Art Museum of Greater Lafayette focuses on Indiana artists and 19th- and 20th-century American art. Finally, the Indiana University Art Museum in Bloomington, set in an I. M. Pei–designed facility, exhibits works from nearly every culture and time period.

Opposite page, left: Indianapolis Museum of Art. Opposite page, right: Students in the Kuttner Quartet at the Indiana University Jacobs School of Music in Bloomington. This page: The *Richard and Billie Lou Wood Deer Fountain* at Indianapolis's Eiteljorg Museum of American Indians and Western Art.

This page: A flight of sweet wines lined up for tasting at Chateau Pomije in Guilford. Opposite page, left: Miss Indiana State Fair, 2007. Opposite page, right: The gaming floor of the French Lick Casino, with a few of its more than 1,345 slot machines.

Performing arts are world-class as well, led by the Indianapolis Symphony Orchestra, founded in 1930. Indiana University's highly regarded Jacobs School of Music in Bloomington hosts noteworthy performances of its own. Touring theater, dance, and musical groups stop at such venues as Indianapolis's Clowes Memorial Hall of Butler University and the Murat Centre, Noblesville's Verizon Wireless Music Center, Merrillville's Star Plaza Theatre, and Wabash's Honeywell Center, among many others. The Indiana Repertory Theatre in Indianapolis is one of the nation's leading professional regional theaters, and community theater thrives all over the state.

Children have cultural choices all their own, starting with the Children's Museum of Indianapolis, which has 11 galleries highlighting the sciences, history, and the arts. Its latest expansion, anticipated by June 2009, will add 37,500 square feet and include a welcome center. Fun and educational experiences also can be had at the Muncie Children's Museum, Fort Wayne's Science Central, and WonderLab Museum of Health, Science, and Technology in Bloomington.

Adults can cash in on fun of a different kind at the state's gaming venues, including five casinos in northwest Indiana on or near Lake Michigan, four to the south along the Ohio River, and one at the historic French Lick Resort in Orange County. The state's two horse tracks, Indiana Downs in Shelbyville and Hoosier Park in Anderson, both added extensive casino gaming to the mix in 2008 for year-round entertainment.

Finally, it wouldn't be Indiana without fairs and festivals. The biggest is the Indiana State Fair in Indianapolis, which spans 17 days in August and draws 800,000 visitors for live entertainment, midway rides, agriculture and education exhibits, and such unusual fare as deep-fried candy bars and deep-fried bananas Foster cheesecake. The 11-day Indiana Black Expo Summer Celebration in July attracts more than 300,000 visitors from around the country to Indianapolis for concerts and talent competitions, health fairs, boxing competitions, and business events and is arguably one of the nation's largest events focused on the African-American experience. The Vintage Indiana Wine and Food Festival, a one-day event in Indianapolis in June, showcases Hoosier wines and specialties of local restaurants, and the Parke County Covered Bridge Festival, scheduled over 10 days in October, puts the spotlight on the western Indiana county's 31 covered bridges through bus tours that leave the Rockville courthouse lawn daily and include stops for shopping. Though the community of Ligonier lost its marshmallow factory in 1997, the four-day Ligonier Marshmallow Festival continues each Labor Day weekend. And for three July days in Whiting, Eastern European heritage is honored with the annual Pierogi Fest, named for dumplings stuffed with meat, cheese, sauerkraut, or potatoes and spotlighted among Oprah Winfrey's summer food festival picks for 2008.

Living in Indiana might mean a fast-paced urban routine, a luxurious lakefront lifestyle, or a laid-back, small-town way of life. No matter where one chooses to settle down in the state, opportunities are nearby for quality primary, secondary, and college education, cultural experiences, outdoor recreation, and family fun. Given the affordable price tag, there's no reason not to move in and live the Indiana life.

This page: Pill containers, one of many plastics products made by Indiana companies such as Evansville-based Berry Plastics. Opposite page: The Toyota Motor Manufacturing Indiana plant in Princeton.

Manufacturing
FABRICATED, FASHIONED, FORGED

Indiana has long been a place where great things are made. For more than a century, manufacturing has been a solid foundation of the state's economy, creating hundreds of thousands of high-paying jobs. In fact, Indiana has, through the years, been among the top states in the nation in the percentage of the workforce employed in manufacturing—in mid 2008, the figure was nearly 17 percent.

That has been both a blessing and a challenge, because it's no secret that American manufacturing has been threatened by overseas competitors and the movement of production activities to offshore locations. Indiana has not been immune to the challenges; yet while its manufacturing employment has declined, its remaining jobs have become more attractive than ever.

The Bureau of Labor Statistics counted 536,600 Indiana manufacturing jobs in June 2008. That's down by about 90,000 jobs from the number before the onset of the 2001 recession—a trend that many have found alarming. But there's another side to the coin: extensive modernization of facilities, education of the workforce, and the adoption of advanced manufacturing strategies have left the sector in a position of solid economic performance.

CHAPTERTWO

According to the Bureau of Economic Analysis (BEA), manufacturing accounted for $62.7 billion of Indiana's economic activity in 2007—about a quarter of the state's gross domestic product and a figure that has grown by nearly $10 billion over the course of a decade. Durable goods represent about two-thirds of that total, led by the motor vehicle sector, primary and fabricated metals, and machinery. On the nondurables side, prominent sectors include chemicals, plastics and rubber, and food products manufacturing.

The manufacturing community has numerous resources that have helped it proceed toward a more technologically advanced, productive, and profitable future. For example, the Indiana Manufacturers Association offers consultation services for a variety of business issues and serves as a voice for the industry in legislative and public policy matters. And at Purdue University, the Center for Advanced Manufacturing brings faculty experts together with Indiana companies to solve manufacturing problems and improve the way those companies operate.

Economic Drivers

There's debate among historians about where exactly the automobile was born, but some say Indiana and point to an early gasoline-powered vehicle created by Elwood Haynes in Kokomo. Whether or not that was the first horseless carriage, it's indisputable that Indiana was the earliest center of automotive manufacturing, before mass production began in Detroit. Today the state remains a key player in the making of motor vehicles and automotive parts, and one whose role is growing. The *2008 Indiana Manufacturers Directory*, published by Manufacturers' News, counted 106,806 Indiana jobs in transportation equipment manufacturing. The BEA pegs the economic output of Indiana's motor vehicle sector at $10.5 billion.

Most obvious are the auto assembly plants, which can be found across the state. In the southwest, for example, is Toyota Motor Manufacturing Indiana, launched in 1996 in the community of Princeton, just north of Evansville. About 4,500 people earn a living at the plant that initially built Tundra pickup trucks, then added Sequoia sport-utility vehicles (SUVs) and Sienna minivans. The plant benefited from several expansion projects that built its annual capacity to 350,000 vehicles, though the downturn in the automotive industry forced the company to shift gears in 2008. Toyota halted domestic production of Tundra pickup trucks in August but vowed not to lay off any workers while revising its manufacturing lineup to better reflect the economic times.

Toyotas also roll off the assembly line northwest of Indianapolis in Lafayette, thanks to a 2006 partnership with Subaru of Indiana Automotive (SIA). Subaru builds its Legacy, Outback, and Tribeca models at its Lafayette assembly plant, which started production in 1989 as a joint venture involving Subaru and Isuzu. Production of Isuzu vehicles in Lafayette ended in 2004, but in 2007 the plant began turning out Toyota's popular Camry, a development that added more than 1,000 new jobs. Through the years, SIA has implemented new manufacturing technologies and processes that have made the Lafayette plant one of the nation's greenest auto assembly plants. In 2004 it became the first of its kind to go "zero landfill," meaning nothing from its manufacturing processes ends up in landfills.

Since 1986 Fort Wayne, in the northeast, has been a major General Motors Corporation (GM) assembly site, with about 2,500 employees turning out Chevrolet Silverado and GMC Sierra pickup trucks at a rate of about 200,000 a year. Though fuel cost increases dampened sales of pickup trucks and caused GM to consider closing some of its pickup manufacturing operations, the Fort Wayne factory was not on the closing list proposed in spring 2008. Another GM brand adversely affected by gas costs is the Hummer, a line of large SUVs originally developed by South Bend–based AM General. The vehicle began as the Humvee (HMMWV) military truck, and in 1992 AM General started production of the Hummer, a civilian version. General Motors bought that brand seven years later and took over marketing and distribution, but in 2008 the company began stepping away from it. Despite the uncertainty facing the civilian Hummer, there continues to be strong demand from the military for AM General's Humvee.

Indiana's newest auto assembly plant is in Greensburg, where production of Honda Civic sedans began in fall 2008. The Honda Manufacturing of Indiana plant was built with an annual capacity of about 200,000 vehicles, and from the start was attracting supplier operations to the area. About 2,000 people are on the payroll at the new plant.

Much larger vehicles emerge from Indiana assembly lines as well. Northern Indiana is a center of recreational-vehicle (RV) production, home to such prominent manufacturers as Coachmen Industries, Jayco, Skyline Corporation, Newmar Corporation, Forest River, and Gulf Stream Coach—all of which reported 2007 sales of $150 million to nearly $600 million. Over-the-road truck trailers are also produced in Indiana, at Wabash National Corporation, headquartered in Lafayette; Vanguard National Trailer Corporation, based in Monon; and Great Dane Trailers, with plants in Brazil and Terre Haute.

Opposite page: Setting up a high-speed milling machine at the Purdue University Center for Advanced Manufacturing. This page, left: Installing rotor assemblies for GM hybrid buses in stator housing at Allison Transmission's Indianapolis plant. This page, right: Enjoying a family road trip with a Jayco RV, made in northern Indiana.

Vehicle assembly is just the tip of the iceberg when it comes to automotive manufacturing. About 133,000 Hoosiers earn their living making transportation equipment and components. One of Indiana's largest and most profitable companies since 2004 has been Cummins, a Columbus-based maker of diesel and natural gas engines and power generators. The company, which recorded 2007 annual sales of more than $13 billion and profits of $739 million, sells its products in 160 countries, including the lucrative emerging market of China, which Cummins first explored in 1975 and where it has had manufacturing operations since 1995. One secret to the company's recent success has been staying ahead of environmental laws and market demands through new technologies that result in cleaner-burning diesel engines. Two-thirds of its 2007 research budget was focused on emission control.

Diesel engines also are one of the mainstays at the Indianapolis operations of Navistar International Corporation. Navistar's International Truck and Engine Corporation plant supplies engines to Ford Motor Company; its nearby foundry, Indianapolis Casting Corporation, casts engine parts for the automaker. Together the plants employ more than 1,000 Hoosiers, and though layoffs caused a drop in that number in spring 2008 as Ford cut back orders to work through an oversupply of pickups, that situation was seen as temporary.

The Indiana landscape is dotted with other manufacturers producing components of cars or trucks. In Kokomo, for example, Delphi Corporation maintains its Electronics & Safety divisional headquarters and employs about 3,500 people. The plant has made car radios since 1935, with earlier models bearing GM's Delco brand. As electronics become more enmeshed with the operations of cars, Delphi is developing new, sophisticated systems that also focus on safety—such as warning systems that alert drivers when a vehicle is in their blind spot.

Chrysler also provides work for more than 5,000 people in Kokomo, where it maintains three modern transmission plants and a metal casting operation. Another transmission plant under construction down the highway in Tipton was announced as a joint venture between Chrysler and German transmission designer Getrag, though by October 2008 the tight credit market had put a question mark beside the plans to open the facility in 2009.

This page, left: Navistar International Corporation's LoneStar truck, whose diesel engine was built at the company's Indianapolis plant. This page, right:

Other auto parts manufacturers with significant operations in Indiana include Aisin USA Manufacturing, a Japanese-owned maker of braking systems, door frames, and other components that provides work for 3,600 people in Seymour, where it is headquartered; Pendleton-based Remy International, a private company that employs more than 6,000 people making starters, alternators, and other electrical components; and General Motors, which employs about 3,500 workers at components facilities in Bedford, Marion, and Indianapolis. Accuride Corporation, a public company headquartered in Evansville, builds numerous automotive components at facilities throughout North America.

Pedal to the Metals and Materials

Indiana has been manufacturing steel since the early 20th century, and through modernization, its mills have endured years of consolidation and foreign competition. The industry today looks significantly different from the one that initially grew up along the shores of Lake Michigan, but the Hoosier State remains a steel powerhouse, marked by new ideas and profitability that some thought wasn't possible in the industry anymore. Steel represents the major part of Indiana's primary metals output, which the BEA tallied at $4.3 billion in 2006 (the most recent figure available).

The most recognizable name in steel continues to be United States Steel Corporation, which built the city of Gary as a company town and today maintains its largest mill there. The Gary Works, spread across 3,000 acres and employing more than 6,000 people, and its sister plants turn out 7.5 million tons of raw steel every year for use in products from appliances to cars and buildings.

The name is not as well known in America as U.S. Steel, but ArcelorMittal, created from a merger in 2004, has an even bigger impact on the northwest Indiana economy, employing 10,000 workers at mills that the relatively new company acquired in Burns Harbor and East Chicago. The world's largest steelmaker, according to the International Iron & Steel Institute, ArcelorMittal in 2007 posted $12.9 billion in U.S. sales. The company also is a joint venture partner with Japan's Nippon Steel Corporation in I/N Kote, a galvanized-steel operation in the Indiana community of New Carlisle. That plant is undergoing a $240 million expansion, slated for completion by 2011, that will add the capacity for an additional 480,000 tons of galvanized steel.

Even as the steelmakers along Lake Michigan were modernizing their giant mills in the 1990s, a new kind of steel industry began to grow elsewhere in the state, sparked by the minimill concept that was pioneered in Crawfordsville. It was there, in 1989, that Nucor Steel constructed its first thin-slab casting facility, turning out steel in formats closer to the needs of the end user. In 2002 the site was the proving ground of another new steel technology, Castrip, which enables thin metal sheets to be cast directly from molten steel.

Steel Dynamics, founded by steel industry veterans in the northeast Indiana town of Butler in 1993, also is built on the minimill concept and, despite foreign competition that has pained the industry in general, is proving that steel can be produced in America very profitably. In its short history, the company, now headquartered in Fort Wayne, has seen its sales grow to nearly $4.4 billion, and in 2007 it turned a profit of approximately $395 million. The company's revenues are expected to grow significantly through the late 2007 acquisition of Fort Wayne–based scrap processor and trader OmniSource Corporation. As a Steel Dynamics subsidiary, OmniSource itself has continued to grow through acquisition, including a June 2008 deal that brought in a South Carolina metal recycler.

Among nondurable goods, chemicals have taken a key position in Indiana's economy. According to the BEA, in 2006 chemicals accounted for 19.7 percent of the state's manufacturing economy, surpassing motor vehicles and parts by 2.5 percent. More than 16 major companies, with sales between $5 million and $2 billion, are based in Indiana, from Elkhart to Evansville. Indianapolis alone is home to eight of them, including BSF Diversified Products, Dow AgroSciences, and Vertellus Specialties, whose pyridine, picolines, and vitamin B3 manufacturing operations, as well as its headquarters, are located in the city.

Indiana's production of plastics is growing too, with an output of $3.4 billion in 2006, according to the BEA, up nearly $400 million over the previous year. Between 1990 and 2005, the state's plastics manufacturing grew at a rate seven times the national average, according to the Indiana Department of Workforce Development. In west-central Indiana, plastics manufacturing represents nearly a fifth of all manufacturing employment.

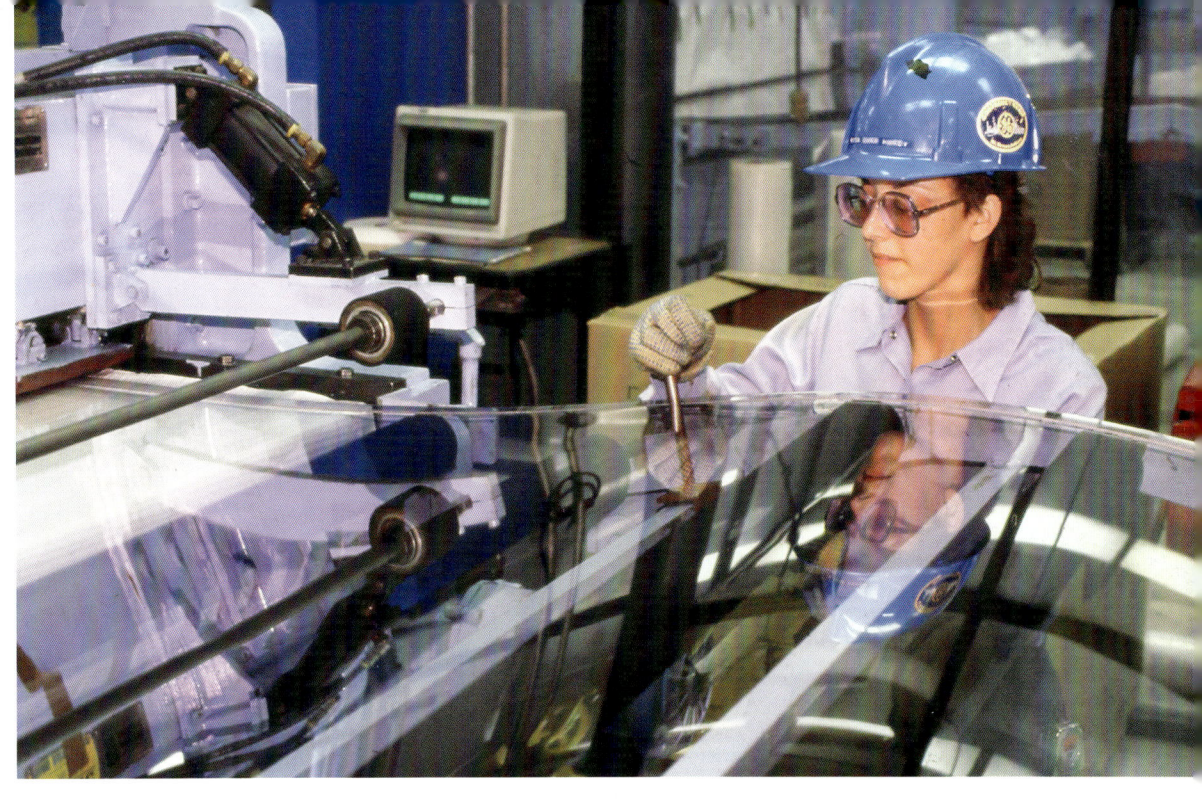

Among the state's leading plastics manufacturers is Evansville-based Berry Plastics Corporation, founded in 1967. A privately owned maker of containers, closures, and housewares, Berry employs about 4,700 people in North America, China, and Europe and reported revenues of $814 million in 2007. The company is growing; in September 2008 it was pledged $3.8 million by the Evansville Redevelopment Commission to expand operations and make Evansville its world headquarters. Berry will add 250 jobs to its operations in the city, more than doubling the current 212 positions.

Not far away in Mount Vernon is SABIC Innovative Plastics, founded as GE Plastics in 1960 to make products from Lexan plastic. The company, sold to the $34 billion Saudi Basic Industries Corporation in 2007, employs more than 1,400 people and stays on the leading edge of the competitive plastics industry by focusing on higher-end, newer plastics, such as Extem, a material that holds up under extreme heat.

Opposite page, left: Coils of coated steel at the United States Steel Corporation's Gary Works. Opposite page, right: A Steel Dynamics employee skimming the zinc pot on the galvanizing line at the company's Flat Roll Division in Butler. This page: Checking the stiffness of a sheet of plastic at SABIC Innovative Plastics in Mount Vernon.

In Terre Haute since 1956, Bemis Company provides work for more than 1,000 people making flexible packaging, such as food packages and soft-drink case wraps. In late 2007, the $3.6 billion Wisconsin-based company saw sales decline and had to cut jobs, but it responded in early 2008 with a program to improve its processes and raise its productivity and was able to bring back many of the employees that had been idled only a few months earlier.

Comforts of Life

Indiana's fame in the furniture store springs from the state's historic abundance of both hardwoods and craftspeople. As Old World as the furniture industry might seem, it has a foot solidly in the future when it comes to products and manufacturing processes. Thanks to ongoing innovation, furniture manufacturing accounts for $1.6 billion of the state's gross domestic product.

Consider Kimball International, an office furnishings industry giant. Based in Jasper, Kimball is the city's biggest employer, with about 4,000 workers there.

The company was launched in 1950 as the Jasper Corporation to build residential furniture and television cabinets. It adopted the Kimball name nine years later, when it acquired the Kimball Piano and Organ Company—a famous maker of pianos—and by the 1960s, the company had grown to become one of the world's biggest piano manufacturers. Today Kimball records nearly $1.3 billion in annual sales and employs more than 7,500 people in the United States and around the world.

Jasper Desk has been making wood office furniture since 1876 in Jasper. The company still mills its own locally grown wood and sells many traditionally styled pieces along with its contemporary variations, yet its record is up-to-date when it comes to the environment. Jasper Desk reuses its manufacturing scrap to heat its wood kilns and buildings and employs processes that limit the emission of harmful chemicals.

Also operating in Jasper is Jofco, a fourth-generation, family-owned firm that employs more than 300 people and is gearing its office furnishings to meet the growing demand for more environmentally friendly products.

Old Hickory Furniture Company upholds Indiana's tradition of outstanding workmanship in Shelbyville, while Swartzendruber Hardwood Creations does so in Goshen.

Kimball International, already mentioned as the office furnishings leader, excels in contract electronics as well, for the industrial, medical, and public safety sectors. Joining Kimball atop Indiana's $1.6 billion electronics manufacturing business are CTS Corporation, an Elkhart maker of electronic components such as automotive sensors, switches, and radio frequency modules that in 2007 recorded sales of $686 million; and Hurco Companies, an Indianapolis-based industrial automation company that makes computer numerical control systems and computerized machines for metalworking, which posted $188 million in 2007 revenues. Lafayette-based Kirby Risk provides electrical and mechanical products and services and had sales in 2007 of more than $317 million.

Indianapolis-based Escient, a division of Digital Networks North America, operates on a smaller scale, but its products are found in some of the country's biggest houses. Escient was founded in 1996 by Indiana entrepreneur Scott Jones, the inventor of the voice mail system used by most of the world's telephone companies. Escient's products help high-end users manage their audio and video home-entertainment needs through such products as centralized servers, in-room control panels, and remote players.

When Indianans are not busy managing their DVD players or iPods, they may be hitting the books. It's hard to find a kid who hasn't read at least one *Harry Potter* book, and though author J. K. Rowling is famously British, many of the books themselves were manufactured in Indiana. R. R. Donnelley & Sons Company employs about 1,900 people in Crawfordsville, printing and publishing about 175 million books annually. None of its products, however, is more famous than the *Potter* books. The printing contract, which called for secrecy, forced Donnelley to place armed guards at the door while it produced *Potter* volumes by the trainload.

R. R. Donnelley is but one player in the state's $1.4 billion printing sector, and the state is prominent in the generation of content, as well. For example, Indiana-based publishing operations create both the *For Dummies* and *Complete Idiot's Guide* series of reference books. The *Dummies* series, from John Wiley & Sons, and the *Idiot's* series, launched by Alpha Books (now a part of Penguin Group USA), are both produced in Hamilton County.

Opposite page, left: The My Desk line of office furniture from Jasper-based Jofco. Opposite page, right: Stylish cabinetry by MasterBrand Cabinets of Jasper. This page, left: A rustic bedroom by Old Hickory, based in Shelbyville. This page, right: Brushing up with the *For Dummies* series, published in Hamilton County.

In the late 1990s, Bloomington emerged as a hub of self-publishing, thanks to Author Solutions, a local company that helps authors get their books onto the marketplace through its divisions, AuthorHouse, iUniverse, and Wordclay. Prior to a September 2007 merger, the company was known as AuthorHouse and was the top American company in the self-publishing business. Its Nebraska-based rival, iUniverse, was number two. The merger united the two under the Author Solutions name and brought iUniverse to Bloomington. Now, with approximately 20,000 new titles in 2008, Author Solutions accounts for one of every 17 titles in the country.

Computer-based and wireless software and services make up another Indiana industry that helps life go more smoothly. One of the companies banking on it is Indianapolis-based Interactive Intelligence, a global firm posting $110 million in sales annually. Interactive Intelligence developed a software product that makes computer- and Internet-based telephone service work on a large scale for call centers and other major corporations.

Another enterprise in this industry is ExactTarget of Indianapolis, which helps companies communicate effectively with customers and prospects through targeted and on-demand e-mail messages. ExactTarget made the Inc. 5000 list of fastest-growing companies in 2008, with annual sales of $48 million and revenue growth of 314 percent between 2004 and 2008. ChaCha, the latest Scott Jones creation, is another—it's a mobile phone service that blends computer search algorithms with input from human guides to provide answers to questions via text message.

Electronic hardware is the order of the day at Sony DADC, where millions of high-definition Blu-ray discs roll off the production lines every month. The company employs 1,300 people in Terre Haute, where it began Blu-ray manufacturing in 2006. Besides being Sony's largest U.S. disc production plant, turning out a total of 60 million compact discs, DVDs, UMDs, and Blue-ray discs each month, the 678,000-square-foot Terre Haute plant also houses Sony DADC's Research and Development facility.

This page: Software by Interactive Intelligence of Indianapolis in use at an Internet-based call center. Opposite page, left: Checking a text response from mobile search service ChaCha, based in Carmel. Opposite page, right: Automated packaging of Blu-ray discs at the Sony DADC plant in Terre Haute.

Despite the headlines of recent years lamenting the decline of American manufacturing, the sector is alive and still quite healthy in Indiana. Companies may have smaller payrolls, but their advanced technologies and processes are keeping them ahead of global competitors and helping Indiana stay solidly on the manufacturing map. The sector has helped Indiana maintain a stronger middle class than many states, and its increasingly advanced nature brings the hope of a bright economic future for the Hoosier State.

This page: Checking plants in an Indiana grain cornfield for insect damage or disease. Opposite page: Newly harvested and threshed soybeans.

Agriculture, Forestry,
and Food Processing

A PLACE TO GROW

It's hard to fathom—most of the Indiana landscape was once forestland. Though thousands of timbered acres remain, especially across the state's southern reaches, those traveling through Indiana's middle and northern sections today will describe an agricultural vista, with corn and bean fields stretching sometimes as far as the eye can see.

Indeed, ever since settlers cleared the land and planted the first crops, Indiana has been widely known as an agricultural state, and its economy and traditions have always been strongly influenced by the farm. Festivals honor Indiana's crops, livestock, farm machinery, and barns. Early-to-bed, early-to-rise farmers have held such sway at the statehouse that Indiana only recently adopted Daylight Saving Time. More intangible but arguably most important, the rural work ethic that grew up on the farm has moved into Indiana's manufacturing sphere, inspiring a culture of quality, loyalty, and productivity.

Agribusiness, of course, starts with the farm. Indiana has 59,000 farms, and the $4.8 billion value of its agricultural products ranks it 15th among the 50 states, according to the U.S. Department of Agriculture's National Agricultural Statistics Service (NASS). Though Indiana today has its share of

CHAPTERTHREE

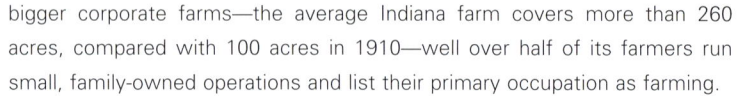

bigger corporate farms—the average Indiana farm covers more than 260 acres, compared with 100 acres in 1910—well over half of its farmers run small, family-owned operations and list their primary occupation as farming.

Indiana is among the nation's top five producers of soybeans, corn, ducks, hogs, and egg-laying chickens. Approximately 5.8 million acres are planted in soybeans and 5.1 million in corn. The state raises 1.1 million ducks a year, 3.5 million hogs and pigs, and 22 million chickens—nurtured as egg-layers rather than for their meat, according to the NASS.

As is the case in many states, agriculture in Indiana is not quite what it used to be. In 1900 the state recorded nearly 22 million acres of farmland, compared to today's 15 million acres. Yet Indiana's economy remains strongly tied to the land. The key to understanding that economy is to realize that agribusiness doesn't stop at the farm gate. It also encompasses a prosperous segment of manufacturing, including the processing of food and other products made from Indiana-grown ingredients, and timber, a crop many people would never associate with Indiana.

Though much of the state has been cleared for farmland, Indiana remains blessed with 43,000 acres of high-quality hardwood forests. This abundant natural resource supports a $1.4 billion wood products industry that provides work for nearly 50,000 Hoosiers, according to BioCrossroads, an Indiana life-sciences support organization. The state is working to establish an Indiana marketing brand for hardwood.

Straight from the Source

Food processing is a significant agribusiness industry in Indiana, and like the wood products industry, it is fueled by the abundance of raw materials, the state's central location (within a day's drive of more than half of the American population), and the presence of expertise. BioCrossroads notes, for example, that the Department of Food Science in the Purdue University College of Agriculture is the largest department of its type in the nation.

According to a 2005 BioCrossroads agribusiness study, the canning and baking segments alone represent annual revenues of approximately $1 billion.

The canning leader is Red Gold, specializing in tomato products. The company was launched in the small Madison County town of Orestes in 1942, during World War II. Founder Grover Hutcherson wanted to support the war effort and invited area farmers to join in by planting tomatoes. The company continued to grow after the war ended, and today, operating from facilities in Orestes, Elwood, Geneva, and Alexandria, Red Gold is Indiana's largest food processor. Dun & Bradstreet estimates the company's annual sales figures at $230 million, which would make it the nation's largest privately owned maker of ketchup and private-label tomato and vegetable juice. Red Gold's approximately 1,100 employees process 50 million cases of tomato products each year.

Indiana also is a leader in the production of popcorn, with a rich history that includes the industry's best-known icon, Orville Redenbacher. The famed bow-tied-and-suspendered image was thought by many to be a marketing fiction, but the spokesman was real—Redenbacher was a native Hoosier and Purdue agriculture graduate who spent decades in Indiana fields developing gourmet popcorn hybrids with his friend and business partner, Charles Bowman. Though Redenbacher's Valparaiso manufacturing facility closed and his popcorn is now manufactured by ConAgra Foods of Omaha, his legacy lives on in Indiana through the annual Valparaiso Popcorn Festival and, even more important, the strength of the industry he championed.

Opposite page, left: Piglets in a high-tech, climate-controlled hog barn. Opposite page, right: Creative advertising for tomatoes processed by the Orestes-based Red Gold company. This page: Cases of Weaver popcorn for delivery to navy ships around the world as part of Operation Popcorn.

The leader of that industry today is Weaver Popcorn Company, founded in 1928 by the Reverend Ira E. Weaver, who sold popcorn in the little town of Van Buren from a horse-drawn wagon. Today the company employs nearly 400 people, mostly in Indiana, and sells to more than 90 countries. Annual sales are estimated by Dun & Bradstreet to be $31.5 million. Though Weaver Popcorn's corporate headquarters has moved to Indianapolis, major operations remain in Van Buren, which now calls itself the popcorn capital of the world and has its own annual popcorn festival.

A well-known name in beans had its origins in Indiana, though the company's headquarters sits in the shadow of the Lucas Oil Stadium in downtown Indianapolis, far from the bean fields. N. K. Hurst Company was launched in 1938 as a family-owned sugar-trading enterprise, and 10 years later made a name for itself in packaged dry beans. Now into its third generation of family ownership, the company packages 18 varieties of beans, including its original trademarked HamBeens 15 Bean Soup.

Few people may recall the Indiana agribusiness connections of the phrase "the greatest thing since sliced bread," still one of the highest compliments that can be paid to a new invention. Though the bread-slicing machine was actually invented in Iowa, the image of sliced bread as a remarkable innovation was nurtured in the 1930s by Continental Baking Company to market Wonder Bread. The product was originally produced by Taggart Baking Company of Indianapolis, which Continental acquired in 1925. A maze of corporate

This page, left: Dried beans, a specialty of Indianapolis-based N. K. Hurst Company. This page, right: A bite of fabulously fresh, whole grain Wonder Bread, baked in Indianapolis. Opposite page: The headquarters of Perfection Bakeries and its flagship Aunt Millie's brand in Fort Wayne.

acquisitions followed, but Missouri-based Interstate Bakeries Corporation continues to bake Wonder Bread and Hostess products in Indianapolis.

Commercial baking also is strong in Fort Wayne, where Aunt Millie's Bakeries, also known as Perfection Bakeries, creates everything from bread and buns to bagels and croutons. The family-owned company, founded in 1901, has 2,000 employees working at seven bakeries in three states and ships its products across the country. Dun & Bradstreet reports company sales of $123.4 million.

Lewis Bakeries aims at consumers' nutritional desires, marketing the Healthy Life brand of baked goods and a half dozen other company-owned brands from its Evansville headquarters. The company was founded in 1925 by three Lewis brothers, who baked bread and delivered it directly to customers' homes. Perhaps its best-known brand through the years has been Bunny Bread. Lewis Bakeries has 2,600 employees in 11 states; Dun & Bradstreet reports its annual revenues at $193.4 million.

Swiss food-processing giant Nestlé has a major, and growing, presence in the Indiana agribusiness industry. One of its long-established operations is in Fort Wayne, where Edy's Grand ice cream and other frozen treats are manufactured in a former Borden facility the company purchased in 1985. The company's Fort Wayne Operations Center grew in 2007–08 as the company boosted its capacity for manufacturing Nestlé Drumstick products and added about five dozen jobs to the more than 300 already there. Another 300 or more Nestlé jobs can be found in Anderson, in a newly opened food-processing operation that produces Nesquik Ready-to-Drink and Nestlé Coffee-Mate products.

This page: Free-roaming hens at Rose Acre Farms in Seymour. Opposite page, left: Automatic milking on the Apple Family Farm in McCordsville. Opposite page, right: Headquarters of pork processor Indiana Packers Corporation in Delphi.

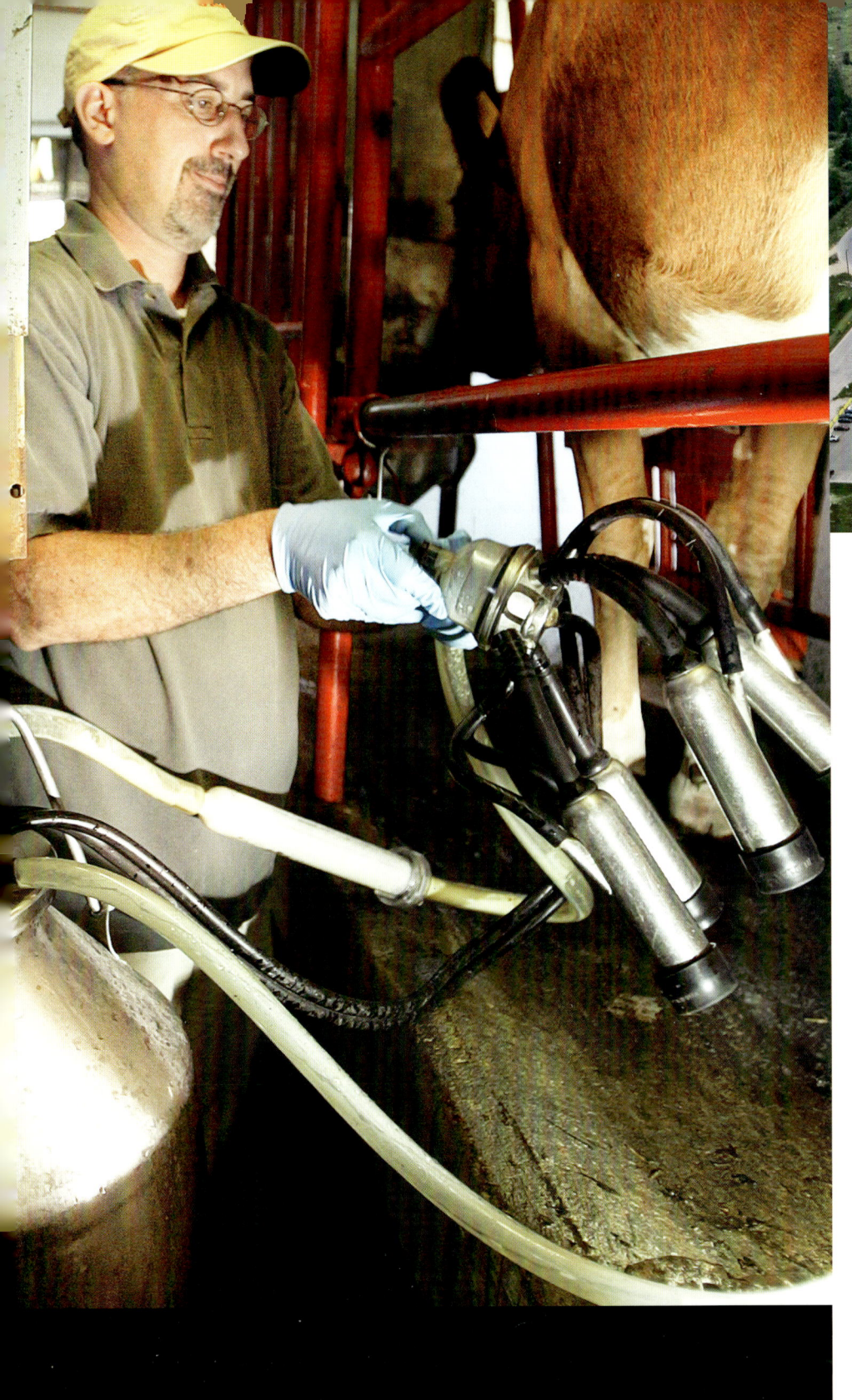

Good Taste

Indiana's vast expanses of farmland are home to numerous livestock and poultry species destined for the dinner table, and the state's pork, poultry, and dairy processors deliver everything from bacon and eggs to Thanksgiving turkeys. BioCrossroads identifies local pork production as an industry ripe for growth, already worth nearly $500 million a year.

The most prominent player in pork is Indiana Packers Corporation in Delphi, whose products are sold under the brand name Indiana Kitchen. Its marketing image evokes the homespun goodness of Hoosier-made, hardwood-smoked pork, but behind the brand is a modern food-processing operation launched in 1991. The Japanese-owned company has over $400 million in annual revenues, according to Dun & Bradstreet, and employs approximately 1,500 people who work in a massive facility that has the capacity to process as many as 14,000 hogs every day into ham, bacon, pork tenderloins, ribs, and related items.

To that bacon, add eggs from Rose Acre Farms, based in Seymour, with annual revenues of about $108 million. The family-owned company has been selling eggs since the 1930s, originally gathering and inspecting each egg by hand, then advancing to an automated operation with millions of hens. Rose Acre Farms now offers buyers the choice of traditional eggs or those from cage-free chickens, along with enriched and organic varieties. The company's approximately 1,300 employees also make dried egg-white powder for use by athletes.

Maryland-based Perdue Farms is one of the leading names in turkey products, and millions of its turkeys are raised in Indiana. The company's operation in the community of Washington employs about 1,100 people and uses turkeys and turkey feed from 140 poultry producers and more than 1,300 grain producers. Every year the plant processes 8.7 million turkeys that collectively weigh more than 200 million pounds. Not far away, in Huntingburg, is Farbest Foods, another prominent producer of turkey products both fresh and frozen.

One of the nation's top duck producers hails from Milford, where more than 1,000 people earn a living at Maple Leaf Farms creating whole, quartered, or processed poultry items that are sold in national chain groceries as well as online. The company, founded in 1958, posts $47.1 million in annual revenues and has a reputation for hormone- and antibiotic-free products, including fresh and precooked duck.

Also in Indiana is food-processing giant Tyson Foods. Its poultry, pork, and baked goods plants in the state include a chicken hatchery and processing facility in Corydon, a chicken-feed mill in Ramsey, a pork processing plant in Logansport, and a Mexican Original operation in Portland, where workers make tortillas, chips, taco shells, and tostadas.

It goes without saying that Hoosiers have raised cows for generations, but a recent trend has given a new twist to the state's dairy industry. Dairy farmers and investors from abroad, such as Dutch-owned Vreba-Hoff Dairy Development, have shown a growing interest in launching dairy operations in Indiana because of the lower cost of land and operations. In the northwest Indiana community of Fair Oaks, for example, Vreba-Hoff supported Bos Family Farms and Fair Oaks Dairy Products, whose Fair Oaks Farms brand can be found on milk and ice cream as well as award-winning handcrafted cheeses.

Support Systems

Indiana's prominence in agriculture has been bolstered by agricultural research conducted at Purdue University in West Lafayette. Departments at the Purdue College of Agriculture specialize in everything from agricultural engineering and economics to botany, forestry, food science, entomology, and animal sciences. Researchers tackle both local and global agricultural issues, such as growing healthier grapes, uncovering more ecological crop management techniques, exploring animal breeding and genetics, and improving biotechnology.

With such expertise readily available, the state has become an attractive location for businesses that serve agriculture. Perhaps the most prominent is Dow AgroSciences, based in Indianapolis, with $3.8 billion in 2007 global sales. Part of the Dow Chemical Company, Dow AgroSciences is involved in a variety of agricultural support activities, including crop protection, pest and weed management, seed development, and agricultural biotechnology. The company maintains a vast research and development operation, containing massive greenhouses where crop varieties and products can be tested, on the northwest side of Indianapolis. Dow AgroSciences also is on the ground floor of the movement to eliminate unhealthy trans fats from the world's diets, creating special canola and sunflower oils.

The significance of Indiana's agricultural heritage cannot be overstated. Its influence can be seen in the landscape and the Hoosier work ethic, and it can be felt strongly in the Indiana economy. Thanks to efforts of civic, organizational, university, and industry leaders, Indiana agribusiness looks forward to a prosperous future.

Opposite page, left: Winner of the 2005 U.S. Champion Cheese Contest—a Swiss-style Emmentaler from Fair Oaks Farms. Opposite page, right: A grocer's display case filled with Tyson chicken items originating in an Indiana hatchery. This page: Researchers in the Indianapolis greenhouses of Dow AgroSciences.

This page: University of Notre Dame cheerleaders during a home game. Opposite page: A Purdue University student working in the chemistry laboratory.

Higher Education

BUILDING TOMORROW'S KNOWLEDGE BASE

Indiana has educated college students since its earliest days as a state, and through the generations, it has developed a collection of educational institutions with reputations that draw students from around the world. The schools range from smaller liberal arts colleges to major research universities and Big Ten campuses. Indiana is home to seven public institutions that operate a total of 40 campuses across the state; its independent colleges and universities number 36. The schools have turned out prominent CEOs, award-winning researchers, thousands of physicians, and even astronauts. And they have boosted the economy by supporting local businesses and creating technologies that have transferred successfully to the marketplace.

Big Contributions in Research and Development

Cutting-edge research at universities across the state is advancing technology, improving lives, and changing the way business is done. Research fields vary widely, from nanotechnology and economics to human sexuality. Leading the way are two public institutions belonging to the Big Ten Conference: Indiana University (IU) and Purdue University. According to the National Science Foundation, Purdue ranked 37th among American universities in 2006 research and development spending, with nearly $373 million in total expenditures. IU ranked 42nd, with $355 million.

CHAPTER FOUR

Indiana University operates eight campuses across the state, serving about 101,700 students—the largest enrollment of any of Indiana's four-year universities. Along with IU's renowned research opportunities, more than 1,000 degree programs draw students from all 50 states and 158 foreign countries. Among the university's best-known programs are business, journalism, informatics, foreign languages, and music. The school was founded in Bloomington in 1820, just four years after the state itself was born. IU Bloomington rolls across 1,933 acres, many of them wooded, and is dotted with dozens of Collegiate Gothic buildings crafted from local Indiana limestone.

IU's tradition of research took off in 1938, when Dr. Alfred Kinsey began his pioneering and sometimes controversial studies into human sexuality that resulted in the creation in 1947 of the Kinsey Institute for Research in Sex, Gender, and Reproduction. The university also was the birthplace of the stannous fluoride formula that would be used in Crest toothpaste, which was tested on 1,200 Bloomington schoolchildren in 1951 and became available in Crest tubes five years later.

The university also operates the Indiana University Medical School (IUMS), based at the Indianapolis campus shared by IU and Purdue: Indiana University–Purdue University Indianapolis (IUPUI). IUMS serves 1,173 students, one of the largest medical school enrollments in the nation. In addition to educating physicians, the faculty performs groundbreaking research. At the Krannert Institute of Cardiology, for example, stem cell and gene therapies are being studied for use in advanced heart failure, and within the division of clinical pharmacology, researchers are studying individualized responses to drug therapy in cancer and psychiatry patients, among others. Top-notch medical care is also available on campus. Champion cyclist Lance Armstrong sought treatment at the IU Medical Center after he was diagnosed with testicular cancer in 1996; in 2005 his foundation established an endowed chair in oncology within the medical school.

But medicine is just the beginning at IU. The school is also a leading research university in the area of computing and related technologies—one of the reasons that IUPUI was chosen to house the operations center for Internet2, an advanced networking consortium of more than 200 American universities working with 70 corporations, 45 public agencies and laboratories, and more than 50 international organizations. The Indianapolis campus is a solid representation of the synergy between university and business: among other things, it hosts the IU Emerging Technologies Center, an incubator for businesses based in life sciences and technology.

Purdue University, with five campuses, is also a statewide system, serving about 72,000 students. It was founded in West Lafayette in 1869 and named for industrialist and benefactor John Purdue. The university is responsible for a wealth of research and technology transfer, particularly involving the companies located at its Purdue Research Park facilities in West Lafayette, Merrillville, New Albany, and Indianapolis. Among the promising stories to emerge from the research parks is that of Endocyte. The company is using Purdue technology to develop cancer-fighting drugs that use the vitamin folate to specifically target cancer cells, which have high numbers of folate receptors. Another company, Imaginestics, developed technology that enables manufacturers to search online for the parts and components they need by sketching them, then uploading their drawings. And yet another, Quadraspec, blends compact disc and laboratory analysis technologies to create a blood test that can detect a host of serious illnesses, including cancer and cardiovascular disease.

Purdue also is betting big on things that are small. Its Birck Nanotechnology Center, completed in 2005, develops technology involving materials as small as a billionth of a meter, applicable to a wide range of disciplines that include agriculture, biology, computer science, engineering, medicine, and physics.

Opposite page, left: Indiana University Bloomington's Indiana Memorial Union. Opposite page, right: Dr. Keith March, a principal investigator on an Indiana University School of Medicine stem-cell study to prevent fat-cell growth. This page: The Campus Center at Indiana University–Purdue University Indianapolis.

Purdue was established as a land-grant university, intended to specialize in agriculture and mechanical arts. To this day, engineering is among Purdue's best-known programs. In 2008 *U.S. News & World Report* ranked five of the engineering programs at the school as among the nation's top 10 such programs, and five more placed in the nation's top 20. Overall, Purdue's engineering program was ranked ninth in the nation. Business education also is a source of pride at the university. Its business program ranked 21st on the *U.S. News* list of top U.S. business programs and departments in 2008. Other well-known fields of study at the school include veterinary medicine, chemistry, interior design, and hospitality and tourism management.

Beyond the Big Ten Schools

Ball State University in Muncie makes noteworthy additions to the state's public educational offerings and research output with undergraduate and pre-professional degrees in 190 areas, plus master's and doctoral degrees in 113. Among its best-known programs are architecture (including landscape architecture), telecommunications, education, and entrepreneurship. With nearly 18,000 students, this public university blends classroom lectures with experiential learning through a range of business connections.

The Entrepreneurship Center, part of Ball State's Miller College of Business, has earned national attention, being ranked by *Entrepreneur* magazine as among the nation's top 20 undergraduate entrepreneurial colleges in 2007. Its "spine sweat" capstone course, called New Venture Creation, requires seniors to create a business plan and present it to a panel of entrepreneurs. If their plan is approved, they pass and graduate; if it is rejected, they must wait until the next year to try again. Through Ball State's Digital Home and

This page, left: Purdue University engineering students modifying an autonomous vehicle they designed and built. This page, right: The annual Wabash River Tube

Healthcare project, part of the university's Center for Media Design, students conduct research for industry partners wanting to learn more about how users interact with the companies' high-tech products. At the center's Institute for Digital Fabrication and Rapid Prototyping, students study ways to manufacture custom products using computer-based design and computer-controlled machinery.

Ball State has been building new facilities filled with hands-on learning technologies, such as the recording studios, control rooms, and editing suites found in the David Letterman Communication and Media Building, christened in 2007 in honor of the late-night talk-show host and Ball State graduate.

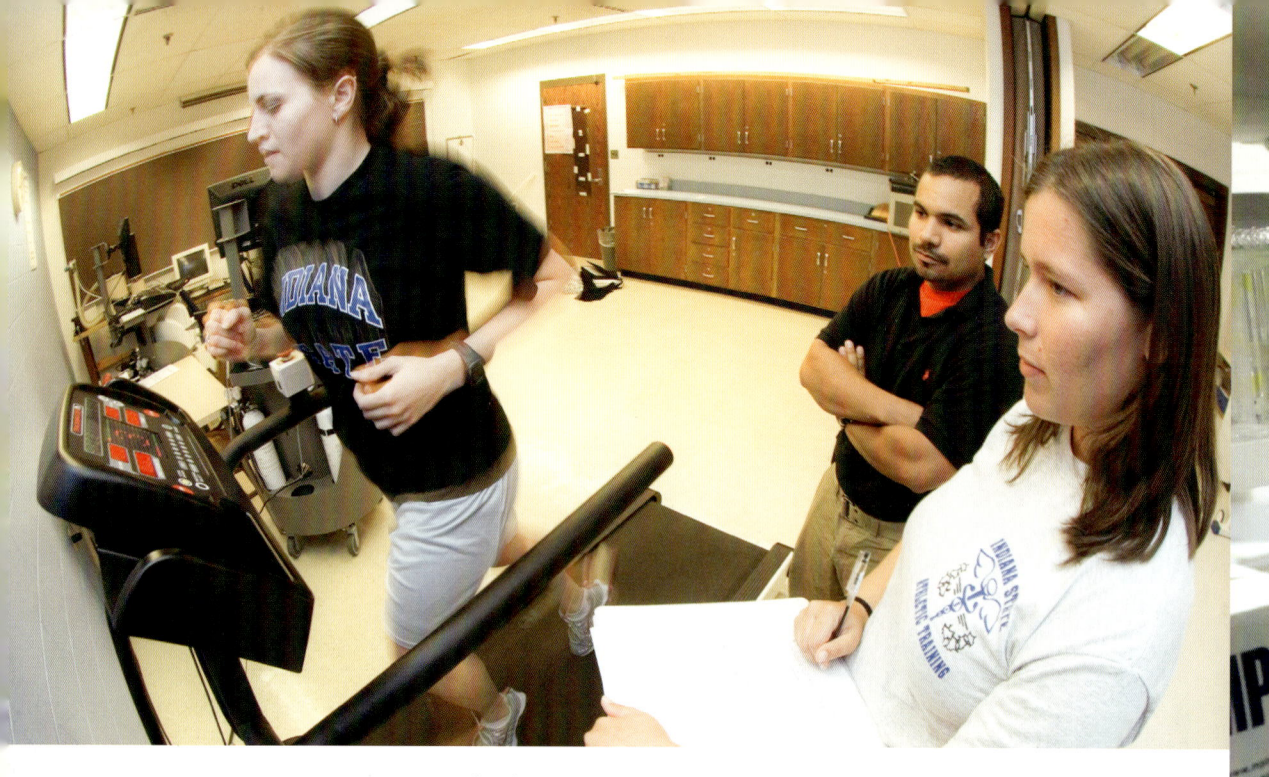

On the western edge of the state in Terre Haute, the public Indiana State University (ISU) serves 10,457 students studying more than 100 major subjects. The most popular fields among undergraduates are early, elementary, and special education; criminology; nursing; business administration; and communication, while graduate students most often choose educational leadership and administration; curriculum, instruction, and media technology; industrial technology education; nursing; and counseling. ISU's strong emphasis on education stems from its founding mission as a teachers' college in 1865.

For two decades, ISU operated a campus in Evansville. That institution gained independence in 1985 as the state-supported University of Southern Indiana, serving 10,126 students—almost as many as its former parent.

Engineering, science, and mathematics are primary focuses at the private Rose-Hulman Institute of Technology, which serves about 1,900 students in Terre Haute. Rose-Hulman carries a solid reputation in scientific fields—among other honors, its engineering program has topped, for nine years, *U.S. News & World Report*'s ranking of the nation's best undergraduate engineering programs at schools where a bachelor's or master's degree is the highest engineering degree offered.

Administrators at Rose-Hulman recognize the value of strong ties to the business world. Along with ISU, the school is part of an economic development and education initiative known as the Terre Haute Innovation Alliance, engaging students and faculty to help local companies develop or expand product lines. And the university's Rose-Hulman Ventures program links students and faculty with technology companies to work in teams designing prototypes, fine-tuning current products, and enhancing capabilities.

Sports fans around the world may be familiar with the University of Notre Dame, which draws about 11,700 students to the South Bend area, but the independent, Catholic institution is not just a sporting dynasty—and not just a fine liberal arts school. It also offers strong programs in business, law, architecture, engineering, and the sciences. Notre Dame has been building its prominence as a research university—since 2000 its annual research funding has doubled to $83 million. Faculty members are conducting promising studies in such areas as aquatic conservation, microfluidics and medical diagnostics, and molecularly engineered materials, to name just a few.

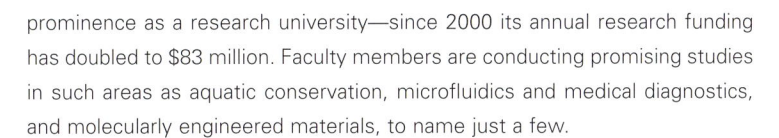

A Traditional Education

Those seeking a traditional, small-school, four-year liberal arts education will find highly regarded institutions such as Hanover College, founded in 1827 on a spectacular site overlooking the Ohio River in Hanover. The school's 1,000 students can either choose their major from among 28 areas of study or design their own. Five centers of excellence at the college focus on business preparation, rivers of the world, free inquiry (an interdisciplinary institute exploring wide-ranging issues), religious leadership, and career development.

Opposite page, left: Athletic training at Indiana State University in Terre Haute. Opposite page, right: Graduate student researchers at the Eck Family Center for Global Health and Infectious Diseases at the University of Notre Dame in South Bend. This page: Hanover College, on the Ohio River.

Two of the state's best-known private colleges—Wabash College in Crawfordsville and DePauw University in Greencastle—maintain one of college football's oldest rivalries. In 1890 the Wabash Little Giants took to the field to play the Tigers of DePauw, and today the teams play for a special trophy, the Monon Bell. Wabash is a liberal arts college for men, serving 900 students from 34 states and more than 20 foreign countries. It grants bachelor of arts degrees; three-quarters of its students typically follow up with graduate studies. The nearly 2,400-student DePauw also sends many of its alumni on to graduate school, and a substantial number have landed prominent positions in business and the arts. Alumni include novelists John Jakes and Barbara Kingsolver, Angie's List founder Angie Hicks, and top executives of such firms as General Mills, Rand McNally, and Cummins.

International connections are prominent at two private institutions, the University of Evansville (UE) and Earlham College in Richmond. UE's total enrollment of 2,577 includes students from 51 countries. The university even offers its own overseas campus, Harlaxton College, located in a manor house in Grantham, England, about 110 miles north of London. At Earlham College, fully 10 percent of the 1,194 students are from other countries.

Career-Focused Learning

A college education has become accessible to a greater number of Hoosiers as the state's primary vocational institution has expanded its offerings. Indiana Vocational Technical College, long known as Ivy Tech, in 2005 formally adopted the name Ivy Tech Community College and added to its liberal arts courses to provide a more well-rounded public education.

Ivy Tech is one of the largest statewide community college systems in the country, with 23 campuses across Indiana and classes in more than 75 communities. Total enrollment in 2008 was 86,130. The system offers easy access, the lowest tuition rates in the state, and a wide range of curricula in eight schools: applied sciences and engineering technology, business, education, fine arts and

This page, left: Wabash College football players with their trophy, the Monon Bell. This page, right: University of Evansville racers after winning NASA's Great Moonbuggy Race at Marshall Space Flight Center in Alabama. Opposite page: Ivy Tech Community College students enjoying a study session.

design, health sciences, liberal arts and sciences, public and social services, and technology. Ivy Tech also creates customized training programs that often are a component of the state's economic development incentives.

Two-year education also is the focus at Indiana's oldest college, Vincennes University, which serves nearly 11,000 students. The public university was founded in 1801 by William Henry Harrison, who would become president of the United States four decades later. Vincennes offers 200 associate degree programs focusing on careers in teaching, technology, and industry, as well as seven baccalaureate programs in areas directly related to Indiana's workforce needs, from health care and homeland security to secondary math education and technology.

Indiana is home to a major national player in career-focused education. ITT Educational Services, headquartered in Carmel, is a public company that operates at least 100 ITT Technical Institutes in more than 30 states. Its curriculum leads to associate's and bachelor's degrees in fields including information technology, electronics technology, drafting and design, health sciences, criminal justice, and business. The company's 54,000 students attend classes year-round on a flexible schedule and have access to online courses.

With its plethora of outstanding colleges and universities, Indiana has been the educational destination for hundreds of thousands of people. The ongoing goal of the state's business and political leaders is to have as many of these graduates as possible continue to put down roots and remain Hoosiers far into the future.

This page: Medical research. Opposite page: The DePuy Orthopaedics line of hip stems.

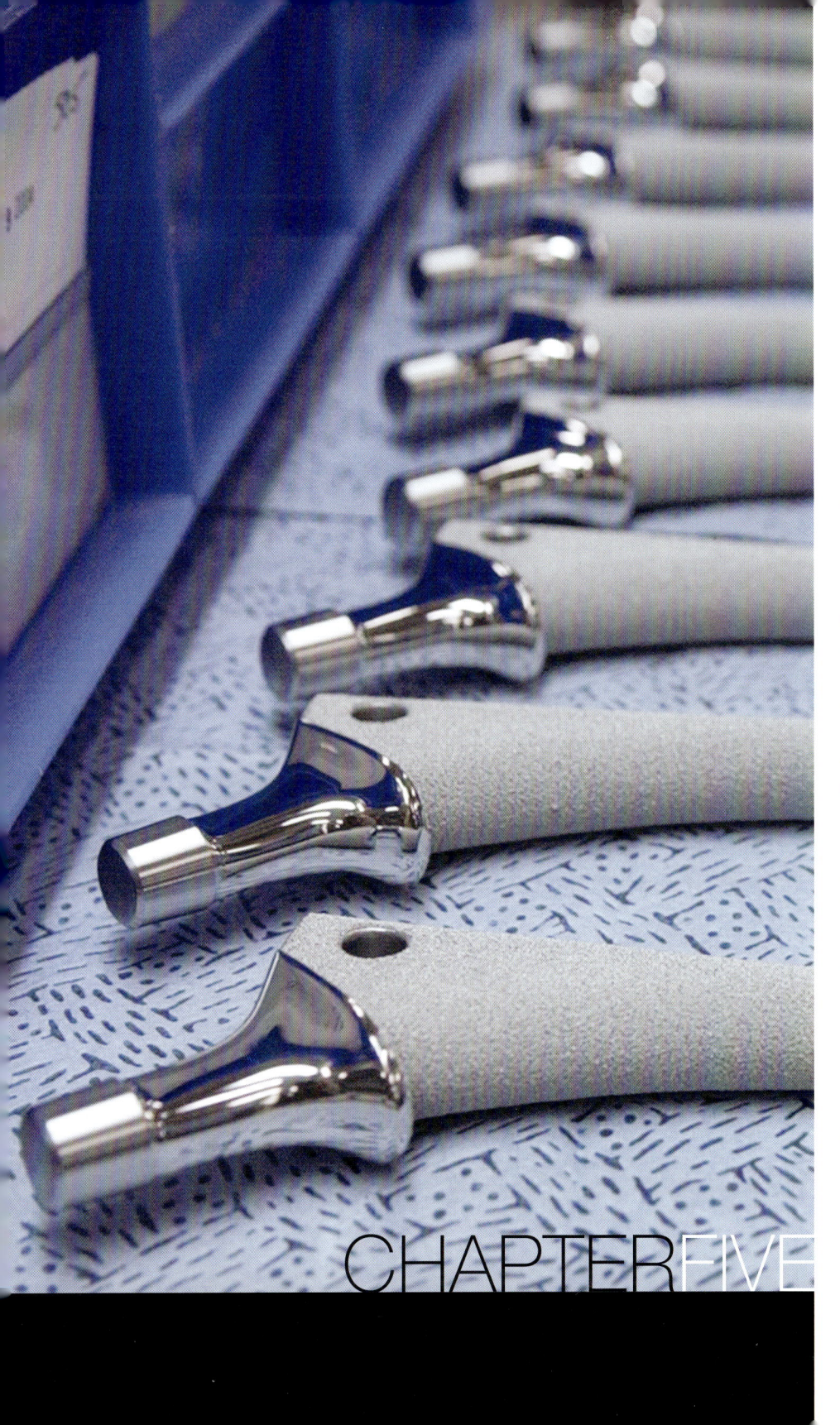

Life Sciences and Health Care

VITAL SIGNS

Indiana is one of the nation's top four life sciences states, according to a study by the Battelle Memorial Institute and the Biotechnology Industry Organization. The report focused on four primary building blocks of the biosciences: drugs and pharmaceuticals; medical devices and equipment; agricultural feedstock and chemicals; and research, testing, and medical laboratories. It found just four states that, in 2006, had strong clusters in at least three of these areas, and Indiana was one of them.

A later study by the same groups measured the impact of life sciences on the economy by metropolitan area and cited Indianapolis as ninth in the nation in total biosciences employment, with approximately 24,000 jobs. In all, more than 275,000 Hoosiers are employed in the business of developing pharmaceuticals, medical devices, and diagnostic equipment, according to BioCrossroads, a group that sponsors initiatives to study and expand the life sciences in Indiana.

These thousands of Hoosiers are seeking and finding treatments for everything from cancer to depression, Alzheimer's disease to sexual dysfunction. They're creating new materials for joint replacements in time to keep aging baby boomers active later in life than their parents were. They're developing products to open clogged arteries and creating new ways to quickly diagnose disease.

CHAPTERFIVE

As with many industry clusters, a variety of circumstances combined to put Indiana near the top of the life sciences charts. Some of the state's earliest companies in this sector began here simply because their founders lived here; that was the case, for example, with pharmaceutical giant Eli Lilly and Company. The founder of Mead Johnson & Company deliberately chose Indiana because of its proximity to the raw materials he needed. Other raw materials—in this instance Indiana hardwoods—bred early success for the company that evolved into hospital bed manufacturer Hill-Rom.

The presence of such companies has helped build life sciences research and education programs at Indiana universities, and that, in turn, has led to the creation and attraction of even more industry players.

Good Economic Medicine

Just how important has the presence of Eli Lilly and Company been to the development of Indiana's life sciences industry? Consider that one of the local venture capital funds working with life sciences companies is named Pearl Street Venture Funds, its name chosen because Pearl Street in downtown Indianapolis is where pharmaceutical chemist and Civil War veteran Eli Lilly first ventured into the business, in 1876.

Eli Lilly and Company's successes through the years have included the first commercial insulin product in 1923, mass-produced penicillin in the 1940s, and the revolutionary antidepressant Prozac in 1988. Some of its more recent blockbusters are depression treatment Cymbalta, the osteoporosis drug Evista, and Cialis, an often-prescribed treatment for erectile dysfunction.

Lilly employs about 40,000 people around the world at its research and development centers, manufacturing plants, and other facilities. With 2007 net sales totaling $18.6 billion and profits of $3 billion, Lilly is the state's second-largest company. Its investments in Indiana include a $1 billion Indianapolis-based biotechnology complex, where the company is researching molecular and cell biology, analytical science, and engineering to develop new biopharmaceutical products. The final phase of the complex was completed in May 2008.

One of the most recognized names in health products is Johnson & Johnson. In 1897 one of the New Jersey company's founding brothers, Edward Mead Johnson Sr., left to begin his own firm, the American Ferment Company. He renamed it Mead Johnson and Company in 1905 and changed its focus to nutritional products for infants. Ten years later, he moved the company to the Indiana city of Evansville to be closer to the source of the ingredients used in his infant formula, Dextri-Maltose.

Mead Johnson now markets 40 product lines, including its best-known line, Enfamil, in more than 50 countries. Though the company became part of what is now Bristol-Myers Squibb in 1967, it maintains its global operations center in Evansville, where the company employs about 1,200 people. In 2008 the parent company announced plans to sell up to 20 percent of Mead Johnson through an initial public offering.

Elkhart earned its place on the life sciences map with the help of Dr. Franklin Miles. His Dr. Miles Medical Company, formed in 1884, started in the business of making tonics and liver pills but really made its mark in 1931, when it introduced the pain reliever Alka-Seltzer to the world. The company became known as Miles Laboratories, and by Alka-Seltzer's 50th anniversary, two billion tablets had been produced. In 1978 the German chemical and health care company Bayer acquired Miles, and Elkhart emerged as the manufacturing home of such Miles and Bayer products as One-a-Day and Flintstones vitamins, as well as Bayer diabetes care products.

In 2004 Cook Pharmica, a contract pharmaceuticals manufacturer, was founded in Bloomington. The company serves the needs of smaller pharmaceutical companies that lack the capital to invest in a manufacturing operation. In its short existence, Cook Pharmica has experienced healthy growth, announcing in 2008 the launch of an $80 million expansion, expected to be ready in 2010, which will add space for vial filling and finishing and formulation development services and add about 200 jobs.

Opposite page, left: Researchers in one of Eli Lilly and Company's Indianapolis laboratories. Opposite page, right: Cans of Enfamil infant formula, developed by Mead Johnson and Company of Evansville. This page: Free fall with fizz for an Alka-Seltzer tablet, made by Bayer in Elkhart.

Healthy Prognosis

Cook Pharmica is just the latest venture to come from the life sciences empire of Bill and Gayle Cook, who launched their first business in 1963 assembling wire guides, needles, and catheters. Their company, Cook, quickly expanded into the manufacture of a variety of other medical products, including stents and tissue-repair products. Its success spawned forays into other businesses as well, including hypodermic needle tubing, plastics processing, and contract pharmaceuticals.

Sprawling across a campus on the northeast side of Indianapolis is Roche Diagnostics, the U.S. headquarters of Swiss-based life sciences giant the Roche Group, which entered the Hoosier State in 1998 through the acquisition of the diagnostics and pharmaceuticals group then known as Boehringer Mannheim. Roche Diagnostics' Indianapolis campus, which develops and manufactures products such as Accu-Chek blood-sugar monitoring and CoaguChek anticoagulation self-testing systems, employs more than 3,900 people. The company's Indianapolis presence has grown to the north of its campus, into the suburb of Fishers, where it expanded its research, development, and production capacity.

This page, left: Cook Pharmica headquarters in Bloomington. This page, right: Roche Diagnostics' AmpliChip test, being used to analyze variations in genes. Opposite page, left: Eli Lilly and Company's Cymbalta capsules. Opposite page, right: Biomet headquarters in Warsaw.

About a century before Roche Diagnostics came to Indiana, Revra DePuy was a traveling pharmaceutical salesman who had a new idea about how to heal fractures. At the time, broken bones were held immobile by hardwood splints, but DePuy thought a fiber-based splint would be more comfortable. That was DePuy Manufacturing's first product, in 1895, and by the late 1960s, the company was venturing into the emerging field of joint replacement. Today DePuy has four business segments: hip and knee replacements, spinal diagnostics, sports medicine and soft tissue repair, and neurosurgical tools. The global company is headquartered on appropriately named Orthopaedic Drive in Warsaw, where 1,300 of its 5,600 employees work. DePuy recorded 2007 sales of nearly $393 million. In 2008 the newest technologies to emerge from its orthopaedics segment were an anterior approach to total hip replacement, the next generation of software for computer-assisted knee and hip replacement, and an advanced line of high-performance knee-replacement instruments.

The event that solidified Warsaw's place as an orthopedics center took place in 1926, when DePuy national sales manager Justin Zimmer resigned and started a competing company to make what he believed was a better splint. His company, like DePuy, moved into joint replacement, and today Zimmer, too, is recognized as a world leader in the orthopedics business. In 2007 it employed 7,600 people and posted revenues of $3.9 billion.

A later entry into the orthopedics business was that of Biomet, established in 1977 in Warsaw. The company was led for almost 30 years by cofounder Dane Miller, who previously had been involved in orthopedic research management at Bristol-Myers Squibb and Bayer. Biomet's original niche was creating especially durable orthopedic implants, and the company was an early adopter of titanium as the alloy of choice. Its products today include those implants as well as a range of other surgical and nonsurgical products, including bone cements, electric bone-growth stimulators, and dental implants. By fiscal 2007, the company employed 6,500 people around the world and its sales had grown to $2.1 billion.

Another small Indiana town landed on the life sciences map in the late 1920s through the efforts of the Hillenbrand family of Batesville, owner of a local casket company. William A. Hillenbrand spearheaded a project to launch a community hospital, and within two years his casket business was supplemented by the manufacture of wooden hospital furniture. That venture today is known as Hill-Rom Holdings, with product lines that include patient beds, stretchers, and a hospital communication system. The company has annual revenues of $1.5 billion and employs about 6,500 people globally, but it still maintains its headquarters in Batesville.

Research and Development Corridor

Interstate 65 between West Lafayette and Indianapolis is lined with cornfields, and State Road 37 between Indianapolis and Bloomington goes through lovely, wooded hills. The landscape may not reveal it, but these highways form the nexus of a growing life sciences corridor.

At Purdue University in West Lafayette, Discovery Park and Purdue Research Park are hotbeds of advancement in a variety of scientific fields, including the life sciences. Purdue Research Park has been helping bring ideas from the laboratories to the marketplace since the late 1990s, while Discovery Park was launched in 2001. The latter's 40 acres of facilities include centers for the biosciences, health care engineering, and oncology. One of the major components of Discovery Park is the Alfred Mann Institute for Biomedical Development, created and endowed in 2007 with a $100 million gift from the Mann Foundation for Biomedical Engineering. The institute is housed in the new Gerald D. and Edna E. Mann Hall, a facility for the park's research centers.

Among Purdue's other programs aimed at cultivating life sciences expertise is the Purdue University Interdisciplinary Life Science Ph.D. program, or PULSe, which puts graduate students through different laboratory rotations to help them choose a research area.

In Bloomington, the Indiana Life Sciences Initiative brings together the diverse academic and research assets and business development expertise available at Indiana University. The Center for the Business of Life Sciences was launched at IU's Kelley School of Business in 2005, along with a new executive certificate program. The idea is to help scientists and researchers learn the business behind the life sciences, so they'll be better equipped to commercialize their discoveries. New on the Bloomington campus since 2007 is Simon Hall, a multidisciplinary science building designed to bring together researchers in such areas as genomics, proteomics, and bioinformatics.

North of Bloomington, the Indiana University–Purdue University Indianapolis campus is another hub of life sciences activity. Beyond the IU School of Medicine and the traditional classroom buildings on the campus itself, IU is surrounding the northern end of the downtown Indianapolis canal with life sciences facilities—a development sometimes called the Bio Opportunity Zone. Among the facilities there are the Indiana University Research and Technology Corporation, which provides technology transfer services, and the Indiana University Emerging Technologies Center (IUETC), a business incubator. IUETC's 25 tenants in 2008 included such companies as Fast Diagnostics, which was working on a system for rapid, accurate diagnosis of kidney function; ImmuneWorks, a biotech company focusing on pulmonary conditions; and PowerHouse Proteomic Systems, working to provide researchers with methods to monitor proteins in living cells over a period of time.

Also based along the canal is the American College of Sports Medicine, a prominent sports medicine organization whose national center has been in Indianapolis since 1984. The more than 20,000 association members around the world are involved in the diagnosis, treatment, and prevention of sports-related injuries, as well as studies in the science of exercise.

Opposite page, left: Researchers analyzing molecule samples in Discovery Park's Bindley Science Center at Purdue University. Opposite page, right: The Purdue Biomedical Engineering Building. This page: An Indiana University–Bloomington scientist with a Tesla magnet, used to study large biological molecules.

This page: A Lifeline helicopter, part of Indianapolis-based Clarian Health's 24-hour critical care transport service, on its way to a local hospital. Opposite page, left: An upbeat pediatric examination at Indianapolis's Riley Hospital for Children. Opposite page, right: Methodist Hospital in Indianapolis.

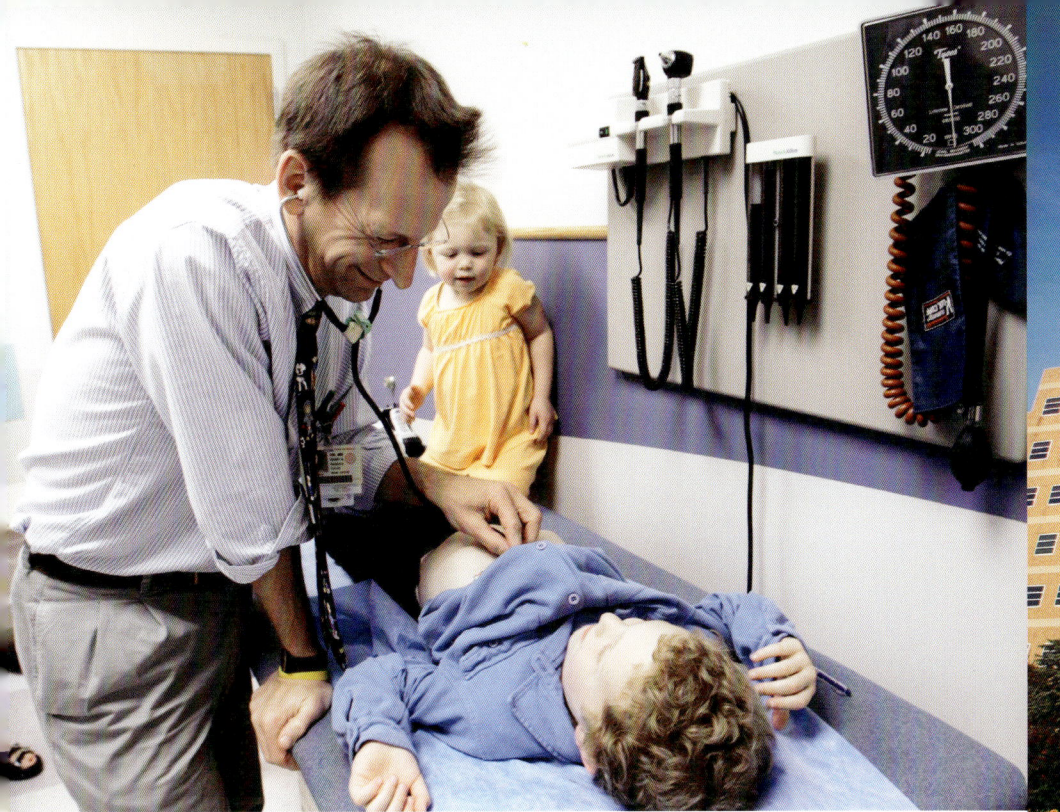

Caring and Curing

Indiana is served by a competitive field of mostly not-for-profit health care organizations, consisting of many small-town hospitals along with several growing, integrated provider groups that operate hospitals and offer other services. The largest such group is Indianapolis-based Clarian Health, originally formed in 1997 as a partnership among Methodist Hospital, Indiana University Hospital, and Riley Hospital for Children. Clarian has since partnered with smaller hospitals across the state and added its own for-profit hospitals in Avon, Carmel, and most recently Lafayette, where the Clarian Arnett Hospital opened in late 2008. Clarian, which staffs more than 1,600 hospital beds, is planning additional facilities across central Indiana.

Indianapolis-based St. Vincent Health, the next largest provider group, is part of the Catholic health care organization Ascension Health. It operates more than 1,500 beds in a network of 17 facilities across the state, led, in terms of bed count, by St. Vincent Indianapolis Hospital, Peyton Manning Children's Hospital in Indianapolis, and St. Vincent Heart Center of Indiana in Carmel. St. Vincent Health serves 45 counties.

The state's third-largest health care system is Indianapolis-based Community Health Network, a group of five central Indiana hospitals—Indiana Heart Hospital and four Community Hospital locations in Indianapolis and Anderson—with staffed beds totaling more than 1,000. The network also operates 70 physician care sites, as well as surgery centers across central Indiana and in Michigan, urgent care clinics, and home health services.

Even in a time of economic turmoil, the life sciences sector is seen as a vibrant engine of job growth and prosperity, which is why virtually every state wants as big a piece of the industry as possible. With its history and critical mass of life sciences companies and research operations, Indiana is well positioned to remain on the leading edge of the business of health.

This page: A Celadon truck arriving in Indianapolis at daybreak. Opposite page: A Republic Airlines jetliner, flying for U.S. Airways Express, landing at Indianapolis International Airport.

Logistics, Transportation, and Energy
LINKED TO THE WORLD

Hoosiers like to refer to their state as the "Crossroads of America"—and the slogan would be quite fitting, if not for the fact that Indiana also has solid links to the world. The state's strong infrastructure features not only great roads but also excellent rail, air, and water transportation systems. With these assets and Indiana's prime Heartland location, the state has a starring role in the nation's logistics and distribution systems.

A third of the U.S. population lives within a day's drive of Indiana, and each year more than 700 million tons of commercial freight travel through the state. Indiana has eight interstate highways, three maritime ports, four international airports, and 4,200 miles of Class I railroad tracks. More than a quarter million Hoosiers are employed by logistics operations and about 32,000 work in long-distance trucking.

Indiana also is a growing player in energy, serving both new technologies and old. The state witnessed the birth of one of the first and most powerful oil companies toward the end of the 19th century and today is home to an increasing number of wind power, ethanol, and biodiesel production facilities.

CHAPTERSIX

Indiana highways lead in every direction except southwest (though in the coming years, Interstate 69 is to be extended from Indianapolis to Evansville and on to the Mexican border). Traversing these highways are numerous Indiana trucking companies, and one of the state's most prominent carriers is Indianapolis-based Celadon Group. In business since 1985, Celadon, a public company, has grown into an international carrier, delivering goods across North America. The company operates about 2,900 tractors and more than 8,000 trailers, and its approximately 4,000 employees generated sales of nearly $566 million for the company in 2008. Celadon also operates TruckersB2B, a group purchasing cooperative.

Delivering the Goods

Indiana's prime location has long been an asset to the economy. Auto parts manufacturers set up shop here to be near the giant assembly plants of the Great Lakes region. Steel mills thrive because they are situated close to end users, and raw material shipments reach them easily via Lake Michigan. The state's status as a magnet for logistics and distribution companies has exploded in the past 10 to 15 years, with bulk warehouses growing like wildflowers along the interstate highways.

Consider Hendricks County, just to the west of Indianapolis. In the mid 1990s, there were almost no distribution warehouses. Now warehouse space in the county's Plainfield area is measured in the tens of millions of square feet, conveniently located along interstates 70 and 74. The same phenomenon has been taking shape along Interstate 65 to the northwest of Indianapolis, in and around Lebanon in Boone County, and to the south in Greenwood, in Johnson County. Warehouse construction has continued within Indianapolis as well, especially along I-70 near Indianapolis International Airport. In 2007 alone, 5.3 million square feet of modern bulk warehouse space was built in central Indiana, according to data from commercial real estate firm Colliers Turley Martin Tucker.

Indiana is home to several prominent names in the moving industry, as well. A group of 33 moving companies formed a partnership in 1948 to create Atlas Van Lines in Chicago. The company moved its headquarters to Evansville in 1960 and today is the key subsidiary of Evansville-based holding company Atlas World Group, whose equipment and warehouses are owned by approximately five dozen private agents. Atlas World Group's revenues were nearly $986 million in 2007. North American Van Lines, established in Columbus, Ohio, also was launched by moving company agents hoping to create a national network. That was in 1933. In 1947 the company was moved to Fort Wayne, where it remains today. But it is now part of SIRVA, one of the world's largest global relocation and moving services companies. Another recognizable name in moving is Indianapolis-based American Red Ball Transit, founded in 1919. The company posted $28 million in 2007 sales.

A logistics operation of a different sort is Langham, a third-party logistics provider that coordinates both domestic and international transportation, warehousing, freight management, and logistics services for more than 1,300 customers from its Indianapolis headquarters. The company, launched in 1993, bears the family name of the three siblings who established it.

Opposite page: Highway signs along U.S. Route 35 near Fairmount directing motorists to Interstate 69, the "NAFTA Superhighway." This page: Atlas Van Lines

This page: Civic Plaza, part of the new passenger terminal at Indianapolis International Airport. Opposite page: A FedEx Express jet about to touch down at the company's Indianapolis hub.

Large volumes of cargo also move through Indiana by air, thanks in large part to the presence of FedEx, which operates its second-biggest hub at Indianapolis International Airport. FedEx is one reason the airport ranks as the nation's sixth-largest cargo airport, according to the Federal Aviation Administration, handling 2.6 million tons of cargo annually and counting. FedEx has undertaken a major expansion of its Indianapolis operations, which will eventually increase the hub by more than 600,000 square feet. Phase one expanded FedEx's main cargo handling and apron parking areas. Phases two and three will see an aircraft maintenance hangar, a building for ground support equipment, and an additional sorting building. One goal of the expansion is to boost capacity from 75,000 packages per hour to 99,000. The company's airport operation is positively bustling when most people are sleeping, as cargo jets deliver tens of thousands of urgent packages to Indianapolis each night for sorting, then depart again to fly them to their destinations.

Also expanding are Indianapolis International Airport's passenger capabilities. A new midfield terminal, opened in late 2008, is an architecturally stunning replacement for the old terminal, which sat at the eastern end of the airport property and caused long and fuel-consuming aircraft taxiing. Joining the new terminal in the middle of the field between the two main runways are a large parking structure and hotel facilities.

Headquartered at the airport is an airline that many passengers have flown without knowing its name: Republic Airways. One of the biggest airlines of its kind, Republic is somewhat anonymous because it shuttles passengers for major carriers, and its planes typically are painted in the colors of those airlines. Named *Air Transport World*'s 2008 Regional Airline of the Year, Republic got its start as Chautauqua Airlines in 1973 in New York and moved to Indianapolis in 1994, when it increased its business routes for US Airways. Today the company flies smaller jets for five major airlines under partner brand names, such as AmericanConnection, Continental Express, Delta Connection, United Express, and US Airways Express. It operates 230 regional jets and flies to 110 cities in North America and the Caribbean, and it's expanding: in 2008 it added Midwest Airlines as a partner, and in 2009, it will place another 29 jets into service. With 5,000 employees, Republic reported 2007 operating revenues of $1.3 billion. Its financial health is boosted by the fact that its major-airline partners, not Republic itself, bear the cost of fuel.

Outside Indianapolis, three additional international facilities—Gary/Chicago International Airport, Fort Wayne International Airport, and Terre Haute International Airport–Hulman Field—as well as 107 smaller public airports add to the state's passenger travel and air cargo options.

Being a centrally located state means being crisscrossed by railroad tracks—4,165 miles of freight tracks, according to the Association of American Railroads. The majority of those tracks are operated by major national rail carriers such as CSX Transportation and Norfolk-Southern Corporation, but some are the property of shorter-line railroads, such as the Indiana Rail Road Company. Founded in 1986, this 500-mile regional freight railroad hauls more than 15 million tons of goods each year, ranging from coal and chemicals to lumber and refrigerators, to destinations such as Indianapolis, Chicago, Louisville, and Newton (Illinois). A wide range of other rail companies, including the Indiana Southern Railroad, Louisville & Indiana Railroad, Central Railroad of Indiana, and the Chicago, Fort Wayne & Eastern Railroad, operate shorter lines across the state.

Surprising to many people is the fact that Indiana—a state whose capital is one of the few major American cities *not* built on a navigable waterway—is a significant maritime player. The northern and southern ends of the state are linked to the world via Lake Michigan and the Ohio River, and businesses utilize those water connections through the state's three ports: Port of Indiana–Burns Harbor, on Lake Michigan in Portage; Port of Indiana–Jeffersonville, on the Ohio River across from Louisville; and Port of Indiana–Mount Vernon, on the Ohio River near Evansville—in an area known globally for grain, fertilizer, and coal and petroleum reserves. Together the ports place Indiana 14th in waterborne shipping, with 70 million tons of cargo handled annually, and make a $1.5 billion impact on the state's economy. More than 60 commercial tenants, employing 2,500 people, operate within the state's 2,300 acres of ports.

Jeffersonville-based American Commercial Lines also helps Indiana make its mark on the maritime industry. The company is involved in both waterborne shipping and the construction of vessels. Its boats have been plying the inland waterways of the nation since 1915, and today it has a fleet of more than 3,000 barges and 120 towboats. The construction segment of American Commercial Lines, Jeffboat, operates a 68-acre shipbuilding and repair site along the Ohio in Jeffersonville, where it also focuses on research and development. American Commercial Lines employs about 2,900 people, including about 1,000 in manufacturing. Its 2007 sales surpassed the $1 billion mark.

An Energized Future

It was the 1880s, and few Hoosiers realized that man-powered agricultural pursuits were about to give way to energy-powered industry. In March 1886, natural gas was discovered in Portland (Indiana), and before long, profitable wells were being drilled across the east-central part of the state. Just as Pennsylvania gas discoveries had built glass industries around Pittsburgh, the seemingly limitless supply of energy in Indiana brought new prosperity to communities like Muncie, where the Ball brothers crafted glass canning jars and left a rich legacy that includes a university and hospital bearing their name.

Opposite page: A CSX train transporting coal through central Indiana. This page: Men repainting a ship in dry dock for repairs.

Many of the state's major energy distributors today have a hand in both natural gas and electricity. The largest of the Indiana-based utilities is NiSource, a Fortune 500 company headquartered in Merrillville that generates electricity and distributes both electricity and natural gas. Its 3.8 million customers stretch from the Gulf Coast to New England. With 2007 revenues of nearly $8 billion, NiSource is one of Indiana's largest companies. Evansville-based Vectren, another large energy corporation, operates three utilities: Indiana Gas Company, Southern Indiana Gas and Electric Company, and a natural-gas delivery company in Ohio. Vectren delivers natural gas to about 990,000 customers and electricity to 140,000. It posted $2.3 billion in revenues in 2007. And Indianapolis-based ProLiance Energy, established in 1996, supplies natural gas to more than 2,400 customers in 18 states.

Millions of other Indiana energy customers are served by companies that have moved into the state through acquisition. Duke Energy, for example, based in North Carolina, has Hoosier customers who once made their payments to Public Service Indiana (gtPSI Energy) and then to Ohio-based Cinergy Corporation. Indiana's largest electric company, Duke delivers energy to more than 780,000 customers in 69 of the state's 92 counties. Indianapolis Power & Light Company, which serves approximately 465,000 customers in the state, is now part of the Virginia-based AES Corporation, which is among the leading independent power-generating companies in the world. Indiana Michigan Power, headquartered in Fort Wayne and now part of Ohio-based American Electric Power, distributes energy to nearly 456,000 Hoosiers.

Across Indiana, the newest ideas in conservation and energy are gaining momentum as Hoosiers become more concerned about the environment. Indiana American Water, for example, an award-winning utility serving more than a million customers in the state, is behind a number of source-water

This page: The BP oil refinery in Whiting. Opposite page, left: NiSource workers marking power distribution points on a grid map. Opposite page, right: Power lines in Lawrence County.

protection programs and works to preserve wildlife habitats and threatened species. Independent developer Orion Energy built a major wind farm across 10,000 acres in Benton County that went online in 2008. Duke Energy signed a 20-year contract with Orion to purchase up to 100 megawatts of its green power. Another wind power developer, enXco, also found a suitably windy site in Benton County and, under a similar 20-year agreement, will be supplying Indianapolis Power and Light customers with energy from the site in 2009.

Biodiesel also is an up-and-coming energy product with strong Indiana roots—soybean roots, that is. Several manufacturing operations across the state are taking Indiana-grown beans and converting them into fuel suitable for powering over-the-road trucks and other diesel-burning engines and generators. For example, in 2006 Integrity Biofuels of Morristown began manufacturing a biodiesel blend that is already seeing $1.3 million in annual revenue. Evergreen Renewables launched a biodiesel plant in Hammond in 2006 that produces five million gallons of fuel annually, and Louis Dreyfus Corporation built an integrated facility in Claypool in 2007 that now produces 88 million gallons a year, ranking it as one of the nation's largest such facilities.

Indiana also has witnessed a rush to build ethanol plants. Among those already producing the corn-based fuel are the Andersons in Clymers, VeraSun Energy Corporation in Linden, Central Indiana Ethanol in Marion, Iroquois Bio-Energy Company in Rensselaer, Poet in Alexandria and Portland, New Energy Corporation in South Bend, and Indiana Bio-Energy in Bluffton. More ethanol-generating facilities are under way in Mount Vernon and Harrisville.

The new biodiesel and ethanol facilities join a longtime producer of gasoline products. John D. Rockefeller's Standard Oil Company incorporated in Indiana in 1889; its refinery in Whiting began operating in 1890. Through the gasoline-burning generations, Standard became Amoco and Amoco became part of BP, but the Whiting Refinery lives on, employing as many as 3,200 workers to process up to 400,000 barrels of crude oil daily. The refinery is the fourth largest in America and the second largest in BP's worldwide network.

The Whiting plant stands as a reminder of Indiana's long history in the business of energy. In energy and transportation alike, Indiana will continue to play a prominent role in powering innovation.

This page: A family enjoying quality time in their new, modestly priced Indiana home. Opposite page: Irwin Financial Corporation's Union Bank in Columbus.

Real Estate, Construction, Finance, and Insurance

BUILDING BLOCKS OF SUCCESS

Every year, thousands of retail businesses around the world are touched by an Indiana connection. So are countless businesses seeking modern office and factory locations. Sports teams, too. They're served by prominent Indiana real estate, development, and construction companies. And they benefit from the protection of Indiana insurance companies.

Strong Foundation

Indiana's real estate market has long reflected the state's conservative nature, and given what's been happening in markets elsewhere, many Hoosiers are thankful. Real estate values in the state typically have risen steadily but slowly—missing the breathtaking boom enjoyed in some places but also spared much of the painful bust.

It's a situation that has always yielded affordable housing. National Association of Realtors' statistics showed a median U.S. home price of $212,400 at midyear 2008, but the median was just $118,400 in Indianapolis and $96,500 in Fort Wayne. Even the most upscale part of Indiana—fast-growing Hamilton County—had a median home price of just $164,340. Housing affordability is one of the drawing cards for Indiana companies trying to lure talent from such places as San Francisco (median home price $684,900) or Boston ($366,100).

CHAPTER SEVEN

Similar things can be said about commercial real estate. Land prices and lease rates did not suffer the spikes seen in some other parts of the country but also have not plunged during economic slowdowns. Office vacancy rates have been relatively predictable. For example, the rate in Indianapolis at midyear in 2008 was 17.2 percent (up from 17 percent at the end of 2007), with enough supply to keep lease rates in check but not so much as to signify economic trouble. Building projects tend to be moving further from the central business districts, following the lead of lower land prices.

Simon Property Group is the most recognizable name in Indiana commercial real estate. Brothers Melvin and Herbert Simon opened their first retail center, Bloomington's Southgate Plaza, in 1960, and by the end of the decade, their Indianapolis-based company, Melvin Simon & Associates, was developing shopping malls at a rate of a million square feet a year. In another decade, the company would begin opening three enclosed malls annually. Among the Simons' many projects, their 1992 Mall of America in Minnesota was an industry head-turner, for its sheer size as well as its enclosed amusement park.

In 1993 the company went public and became Simon Property Group. Through acquisition it grew into the nation's biggest real estate investment trust, according to analyst InRealty. Today Simon owns or has an interest in 383 retail properties, totaling 261 million square feet, across North America, Europe, and Asia and recorded more than $3.6 billion in 2007 sales. Its recent activity in Indiana includes upgrades to University Park Mall in Mishawaka and Greenwood Park Mall in the Indianapolis area, as well as construction of the open-air Hamilton Town Center in Noblesville, just north of Indianapolis.

What Simon is to retail real estate, Duke Realty Corporation of Indianapolis is to office, health care, and industrial property development. The company, launched in 1972, first created the sprawling Park 100 industrial park in Indianapolis. Duke Realty went public in 1993 and grew, through acquisition, to join the ranks of the country's top 20 real estate investment trusts, according to InRealty. Today it controls 142 million square feet of property, along with 7,700 unbuilt acres for future development, in 24 cities across the nation. It also has its own construction company, which is involved in projects both in and away from Duke developments. Among its current Indiana projects are a medical office building in Mishawaka for Saint Joseph's Regional Medical Center, a warehouse at AllPoints Midwest bulk distribution park in Plainfield, and two medical office buildings on Indianapolis's south side for St. Francis Hospital and Health Centers. Duke Realty's sales in 2007 exceeded $1 billion.

Another prominent name in Indiana real estate development is Kite Realty Group Trust of Indianapolis. With nearly $139 million in 2007 sales, the company specializes in retail and office properties and works in Indiana and several other states. In the first quarter of 2008, it had 57 active properties totaling 8.2 million square feet, with 2.2 million more square feet in the works. The University of Notre Dame commissioned the firm to develop Eddy Street Commons, an expansive, mixed-use project adjacent to the South Bend campus; the first phase is slated for completion in autumn 2009.

Lauth Property Group, an Indianapolis developer founded in 1977, works in 35 states. In 2008 the company had 275 projects under way, totaling more than 35 million square feet of office, industrial, health care, and retail space, valued at $1.6 billion. Its Indiana projects have included the TJX Home Goods distribution center in Indianapolis, North Meridian Medical Pavilion in Carmel, and Brownsburg Station shopping center in Brownsburg.

Constructive Uprising

Despite the national downturn in the housing industry, there remains plenty of optimism in the Indiana construction industry. The state has not been immune to mortgage foreclosures, and new-home sales cooled in 2008 as national financing sources dried up. But thanks to growth in other construction sectors, projections for the industry as a whole indicate only a modest drop in contracts—about 1 percent, according to analyst McGraw-Hill Construction.

Opposite page: Featured areas in malls developed by Simon Property Group of Indianapolis: left, the food court at Greenwood Park Mall in Greenwood; right, the play area at the open-air Hamilton Town Center in Noblesville. This page: Duke Realty Corporation headquarters in Indianapolis.

Massive building projects are the specialty of Hunt Construction Group, founded in Indianapolis in 1944. Although Hunt is now headquartered in Arizona with offices across the nation, it maintains its Midwest division in Indianapolis. The company is the top name in sports facilities construction, according to *Engineering News-Record*, but it also is active in just about every other kind of major building construction. Among its local projects are the Lucas Oil Stadium, which opened in downtown Indianapolis in 2008, and Purdue University's Birck Nanotechnology Center in West Lafayette, completed in 2005. Hunt's contracts are valued at up to $8 billion annually.

Brick mason Francis Wilhelm of Indianapolis went into business in 1923 building porches for his neighbors. Today that business, F. A. Wilhelm Construction Company, is one of the most prominent builders in the state, putting its name on numerous big projects including government and university buildings, industrial facilities, hospitals, hotels, and the expanded Indianapolis Museum of Art. It employs an average 1,200 workers of varying specialties and records nearly $266 million in annual sales.

As the name implies, Industrial Contractors, based in Evansville, has plenty of experience in industrial construction, from Duke Energy's Gibson Station in Gibson County to Toyota Motor Manufacturing Indiana projects in Princeton. But the company, in business since 1964, has had a hand in major office projects as well, including the headquarters buildings of public utility Vectren Corporation and Old National Bancorp, both situated along the Ohio River in Evansville. The company's 2007 revenues were $370 million.

Indiana's largest construction companies also include Goshen-based Rieth-Riley Construction Company, which has been specializing in paving and other road projects since 1916; the Hagerman Group, a Fort Wayne–based general contractor established in 1908 that is particularly active in school construction, serving clients such as the University of Notre Dame and Indiana Institute of Technology; and Traylor Brothers, in Evansville since 1946 and focusing on heavy construction, such as bridges and tunnels.

Money Matters

Indiana's financial scene has changed dramatically during the past two decades, as it has across the United States. Many of the state's largest banks merged with even larger out-of-state institutions, offering the business community access to more globally focused financial services but also erasing some long-standing names from the Indianapolis skyline.

This page: Indianapolis Museum of Art expansion, built by F. A. Wilhelm Construction Company. Opposite page, left: Interior of the Indianapolis University–Purdue University–Indianapolis Campus Center, built by Hunt Construction Group. Opposite page, right: An Evansville branch of Fifth Third Bank.

Indeed, the three biggest Indiana banks used to be headquartered within blocks of the landmark State Soldiers' and Sailors' Monument at the center of Indianapolis. Indiana National Bank, whose name once appeared on a downtown Indianapolis skyscraper, vanished through a series of mergers. Parts of it are now known as Regions Bank, and other parts eventually landed in the hands of giant JPMorgan Chase & Company. American Fletcher National Bank & Trust Company's skyscraper was still under construction in the 1980s when the institution was absorbed into Bank One, which later also became part of Chase. And Indianapolis's Merchants National Bank had assembled a collection of smaller Indiana banks of its own before being taken over by National City Bank, which reached across the border into Indiana from Ohio. Another Ohio bank that branched into Indiana—and is now among the biggest names operating in the state—is Fifth Third Bank of Cincinnati.

This page: Headquarters of Old National Bank in downtown Evansville. Opposite page: An exchange at one of the state's long-established banking companies.

Today the largest Indiana-based banking companies are headquartered much further from the capital. The biggest one is Old National Bancorp of Evansville. The institution is nearly as old as the state itself, founded in 1834 and spending much of its life as a community bank, growing slowly. Its first acquisition came in 1985, and from that point on its management pursued a growth-through-acquisition strategy aimed at gaining enough critical mass for the bank to fend off out-of-state suitors. Old National has made more than 40 acquisitions and operates more than 110 banking centers across Indiana, southern Illinois, and western Kentucky. By the end of 2007, the company had $7.8 billion in assets.

Irwin Financial Corporation of Columbus is the state's second-largest bank holding company, with assets of $6.2 billion in late 2007, according to the Federal Deposit Insurance Corporation (FDIC). Its Irwin Union Bank had humble beginnings in the early 1860s, when local businesses began leaving their valuables in the safe at Joseph Ireland Irwin's mercantile store. By the 1990s, it was as big in the mortgage business as it was banking. Irwin sold its mortgage subsidiary in 2005, though it retained some of its mortgage holdings.

To the north, in South Bend, is the state's next-largest institution, 1st Source Corporation, which in mid 2008 tallied its banking assets at almost $4.5 billion. The institution traces its history to 1863—a merger of several local institutions helped the banking company survive the Great Depression. 1st Source has 79 banking centers in Indiana and Michigan, and its Specialty Finance Group operates two dozen locations nationwide.

The trend of larger Indiana banks relocating to smaller cities continues on down the FDIC's list of the biggest bank holding companies: First Merchants Corporation in Muncie, Integra Bank Corporation in Evansville, MainSource Financial Group in Greensburg, First Financial Corporation in Terre Haute, and Lakeland Financial Corporation in Warsaw. The capital doesn't show up on the list until the 13th spot, held by the National Bank of Indianapolis.

Although Indiana-headquartered banks tend not to stray very far across the state line, a different kind of Indiana financial services company serves more than two million customers across the country. American General Financial Services, based in Evansville, provides a wide range of consumer loans, including revolving and installment loans marketed through more than 17,000 retailers hoping to help customers afford their products. Through more than 1,500 branch offices, American General makes other kinds of consumer loans as well, from debt consolidation and vacation loans to mortgages and home equity lines. The company was founded in 1920 as Interstate Finance Corporation to finance auto sales and soon branched into credit insurance to cover borrowers. In 2001 the company became a subsidiary of American International Group (AIG), the insurance giant rescued in September 2008 by the U.S. government. It was not immediately decided whether AIG would retain its ownership stake in American General.

Recognizing the value of venture capital when it comes to new business creation, Indiana continues to strengthen its access to venture capital. The biggest venture capital organization is CID Capital, headquartered in Indianapolis with additional offices in Columbus, Ohio. The firm has raised more than $280 million in seven equity funds and has invested in nearly 70 companies since its 1981 founding. Its major areas of interest include the life sciences, manufacturing technology, information technology, and business services.

Gazelle TechVentures, based in Carmel, was created in 1999 to focus, as its name suggests, on technology investments. It is the fruit of the efforts of government, academia, and the business community, whose leaders collectively wanted the state to offer more venture capital opportunities. The $60 million venture capital organization has supported such Hoosier ventures as Bloomington-based Author Solutions as well as application service provider eTapestry, of Greenfield.

Indiana's venture capital lineup also includes Carmel-based Spring Mill Venture Partners, which backs information technology and early-stage life sciences companies; Indianapolis's Pearl Street Venture Funds, with a life-sciences focus; and Twilight Venture Partners, also based in Indianapolis, focusing on biotechnology and medical technology investments. Other funds include those linked to Indiana organizations such as Eli Lilly and Company, Clarian Health Partners, Purdue University, and Rose-Hulman Institute of Technology.

Sound Protection

Indiana has a long history in the insurance industry, and state officials have made a concerted effort to build upon past successes by streamlining regulatory processes and encouraging the development of insurance-related workforce training and degree programs at state universities. Among Indiana's insurance industry leaders is the state's largest company, WellPoint, which began as an ordinary insurance provider.

There was nothing all that unusual about Blue Cross and Blue Shield of Indiana—at least nothing that would indicate the remarkable growth that was in store. It had been a strong and successful health insurance provider since the 1940s, but it wasn't any different from any other state's Blue Cross and Blue Shield organization. During the 1990s, however, it grew by acquiring other states' plans, and in 2001 the mutual company transitioned into a public stock company known as Anthem. Three years later, it acquired California Blue Cross insurer WellPoint Health Networks, affixed the WellPoint name to its Indianapolis headquarters, and took its place as the nation's largest health insurer. Today WellPoint provides health insurance for 34.8 million Americans—approximately one in nine. The company employs 41,000 people. In 2007 it posted revenues of $61.1 billion and profits of $3.3 billion.

A different type of insurance has made another Indiana company an industry leader. Forethought Financial Group, based in Indianapolis, covers retirement needs and the financial side of prearranged funerals. It was founded in Batesville as an arm of Hillenbrand Industries (the holding company for Batesville Casket Company), with the idea of providing one-stop service to funeral homes and their customers who preplan funerals and buy Forethought life insurance to pay the bills. In 2004 Hillenbrand sold its Forethought unit to the Devlin Group investment firm. Forethought's life insurance operation reported 2007 revenues of about $720 million.

Still another insurance industry leader providing services nationwide from an Indiana base is Fort Wayne's Medical Protective Company. The company was launched in 1899 as the first writer of malpractice insurance for doctors and dentists. In 2005 the firm became part of the Berkshire Hathaway group of companies and now protects more than 70,000 health care providers.

Two of Indiana's most visible insurance companies have their names on downtown Indianapolis landmarks (the OneAmerica Tower and Conseco Fieldhouse, the arena that is home to the NBA Pacers). The first insurer, OneAmerica Financial Partners, had its start in 1899 as American Central Life Insurance Company and, through mergers, became known as American United Life. It grew into a significant player in life insurance and retirement products and in 1982 became the anchor and title tenant of what was then the tallest building in Indianapolis. The company became known as OneAmerica in 2000, and in 2007 it reported assets of $19.9 billion. The other insurer is Conseco Services, a Carmel-based provider of life, health, and retirement insurance products that was founded in 1979 and once sat near the top of the list of largest Indiana public companies. Following a difficult time that included a bankruptcy filing in 2002, Conseco Services has emerged as the state's fifth-largest public company, with annual revenues of about $4 billion.

In real estate, construction, financial services, and insurance, Indiana companies have reflected their home state's conservative mindset—not just a political ideology but a way of doing business that values tradition and careful thought, sometimes at the expense of risk-taking. In many business pursuits, risk-taking is handsomely rewarded, but given today's business climate in these particular fields, a cautious approach has proven to be wise.

Opposite page: The OneAmerica Tower, home to OneAmerica Financial Partners, in downtown Indianapolis. This page, left: Retiring without worry thanks to Forethought Financial Group of Indianapolis. This page, right: Indianapolis's Conseco Fieldhouse, named for insurance provider Conseco Services.

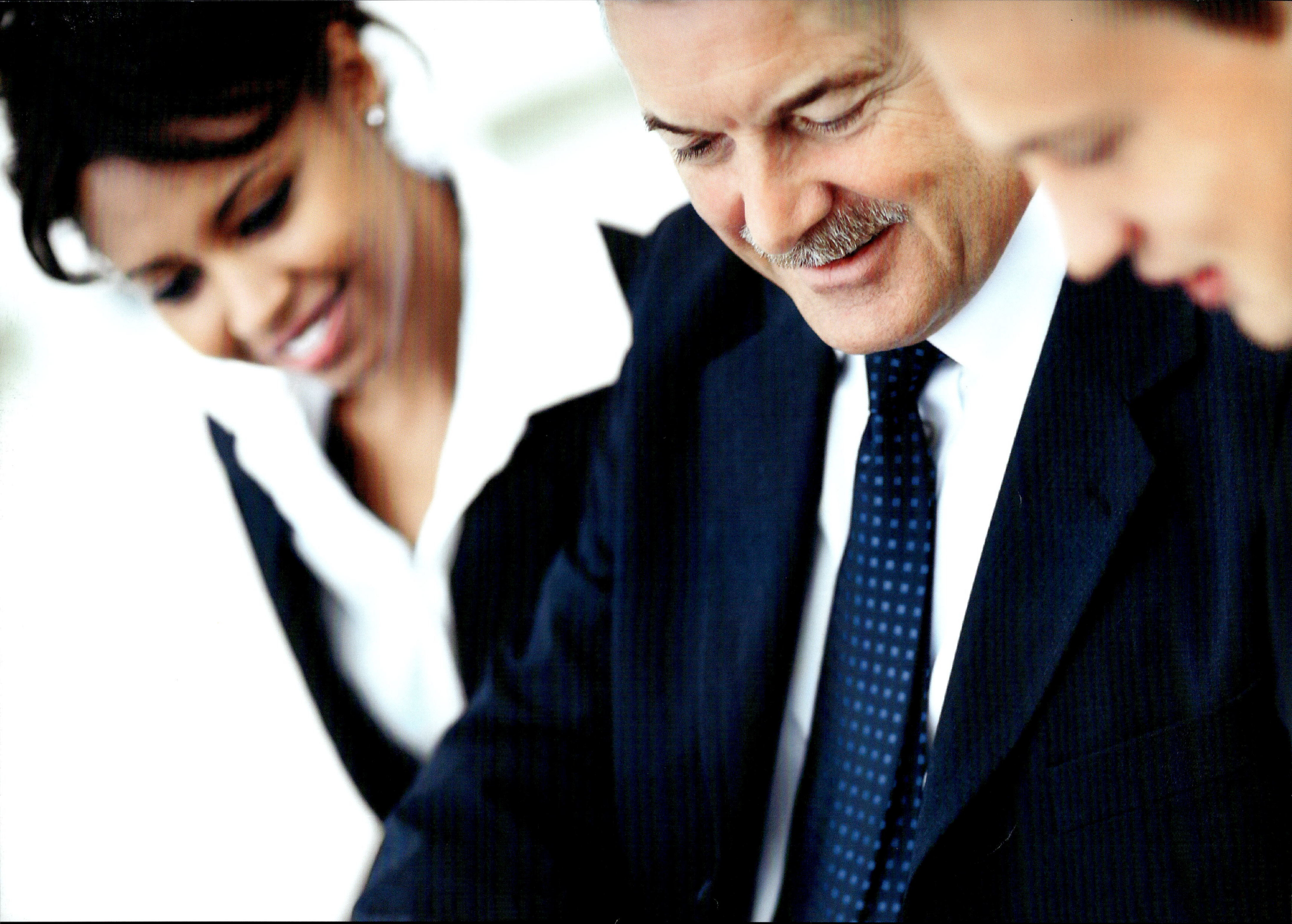

SUCCESS STORIES:
PROFILES OF COMPANIES AND ORGANIZATIONS

Advanced Materials and Technology
Sony DADC, 88–89

Education
Vincennes University, 92–93

Energy, Fuel, and Utilities
BP Products North America, Inc., 96–97
Duke Energy Corporation, 100
Northern Indiana Public Service Company (NIPSCO)—A Division of NiSource Inc., 102

ProLiance Energy, 98–99

Environmental Resources
Indiana American Water, 106–07

Manufacturing and Distribution
Kirby Risk Corporation, 118–19
MasterBrand Cabinets, Inc., 116–17
Old Hickory Furniture Company, 112–13
Steel Dynamics, Inc., 110–11
United States Steel Corporation, 114–15

Vertellus Specialties Inc., 120

Pharmaceuticals, Medical Technology, and Health Care
Eli Lilly and Company, 124–25
Roche Diagnostics, 126

Tourism, Entertainment, and Hospitality
Courtyard by Marriott Indianapolis—Carmel, The, 134
French Lick Resort, 130–31
Hyatt Place Indianapolis Airport, 132
Hyatt Place Indianapolis Keystone, 133

PART TWO

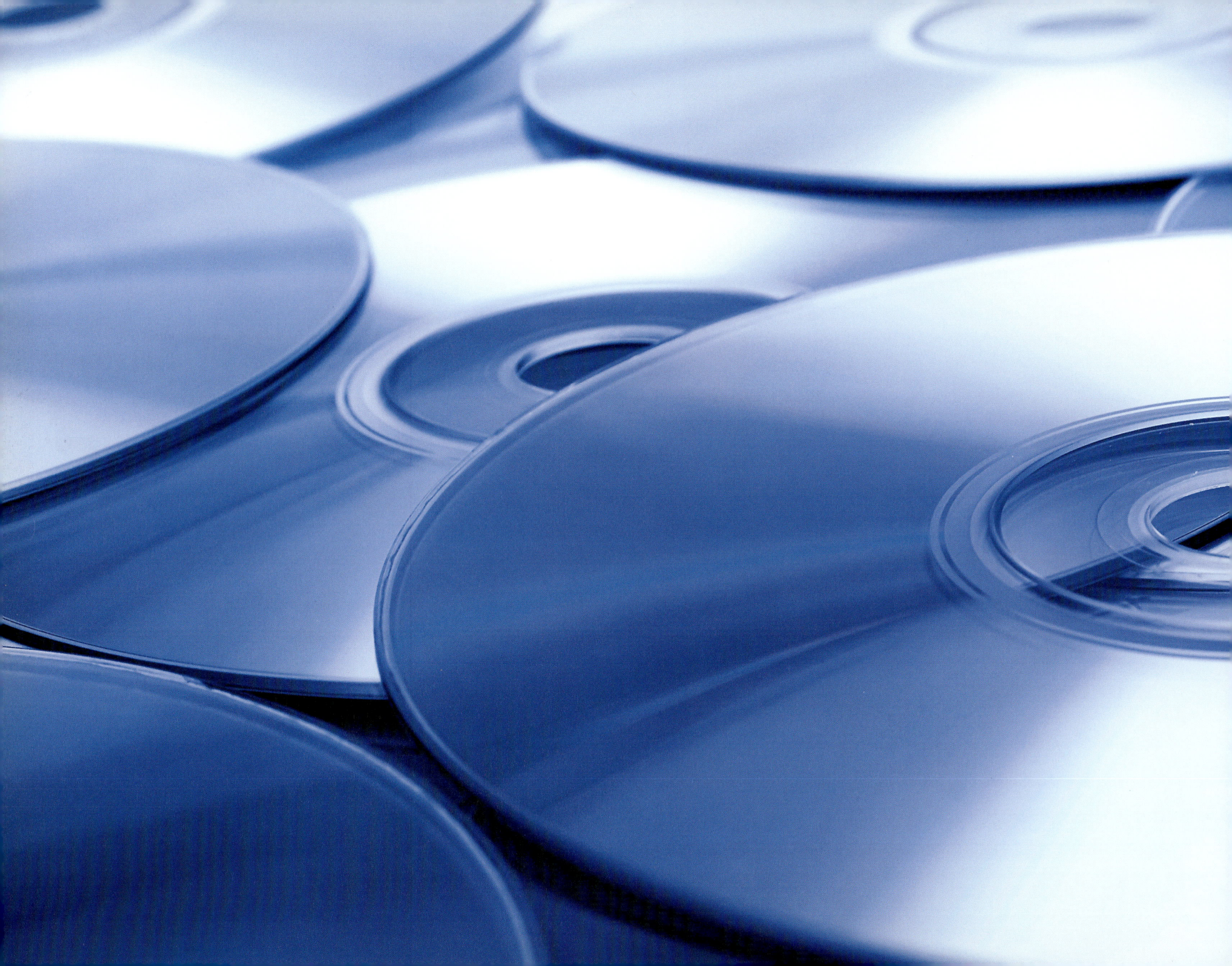

PROFILES OF COMPANIES AND ORGANIZATIONS
Advanced Materials and Technology

Sony DADC

This full-service technology and solution provider delivers digital and physical (CD, DVD, UMD, and Blu-ray Disc) supply-chain service to the most recognizable entertainment and software companies around the globe. For 25 years Sony DADC has operated in its hometown of Terre Haute as a valued employer and community partner.

Above: Sony Corporation's state-of-the-art facility in Terre Haute, Indiana, was the first compact disc plant in North and South America. As of March 2008 the daily capacity at Sony DADC's Terre Haute facility for each format was 700,000 CDs, 1,467,000 DVDs, 500,000 Universal Media Discs (UMDs), and 434,000 Blu-ray Discs™ (BDs).

In 1983 the majority of people in the United States were buying their favorite music either stored on cassette tapes or vinyl records. Sony Corporation, however, had already developed what is known today as a CD (compact disc) at its facility in Japan. The corporation was, at that time, looking for a location outside of Japan for its first manufacturing site for this promising new product.

Sony's search for a manufacturing facility for its virtually unknown CD product brought the company to Indiana, where Sony soon hired an all-Hoosier staff to establish a state-of-the-art manufacturing facility.

Sony not only had faith in its unproven CD technology, but it also had great faith in its growing Indiana team. The facility at Terre Haute was the first CD plant in the United States. Appropriately, the first CD to roll off the line in Terre Haute in 1984 was *Born in the U.S.A.* by world-renowned singer and songwriter Bruce Springsteen. Out of Japanese technology and America's workforce was born a new and popular format that soon experienced tremendous growth across the globe.

CDs: A Spinning Success

Once they hit the marketplace in 1984 CDs quickly began to impact the history of consumer electronics. The $21 million facility in Terre Haute began production with a capacity of 300,000 discs per month and 65 full-time employees. By the end of 1984 the facility had produced two million discs and had grown to 100 full-time employees. The facility's CD-pressing capacity continued to grow—standing at 27 million units produced per month at the turn of this century and with clients ranging from record companies and software developers to schools and libraries.

Originally named Digital Audio Disc Corporation (DADC), today's Sony DADC still retains the acronym as part of its name despite the company's growth into other areas of production beyond audio discs.

New Formats and Continued Expansions

As the flagship facility for Sony DADC in the Americas, Terre Haute is the first site in which new formats are produced in North or South America. In 1997 the Terre Haute facility began production of the DVD, a format used to hold prerecorded movies, music videos, and ROM titles as well as PlayStation 2 game titles. By late 2000 capacity exceeded 10 million units produced per month.

In 2005 the Universal Media Disc (UMD) was introduced, a format specifically created to work with the Sony PlayStation Portable (PSP). The biggest launch in recent years has been the Blu-ray Disc™ (BD), which was launched in May 2006. The Terre Haute facility was the second globally to invest in this new technology—the first being the Sony DADC facility in Japan.

An ongoing expansion in 2008 will nearly double BD capacity to 800,000 discs per day or 23 million per month. This expansion has also increased Hoosier employment—more than 125 jobs have been added in 2008 to handle the increased capacity. This brings the

Photo: © Hickman Photography

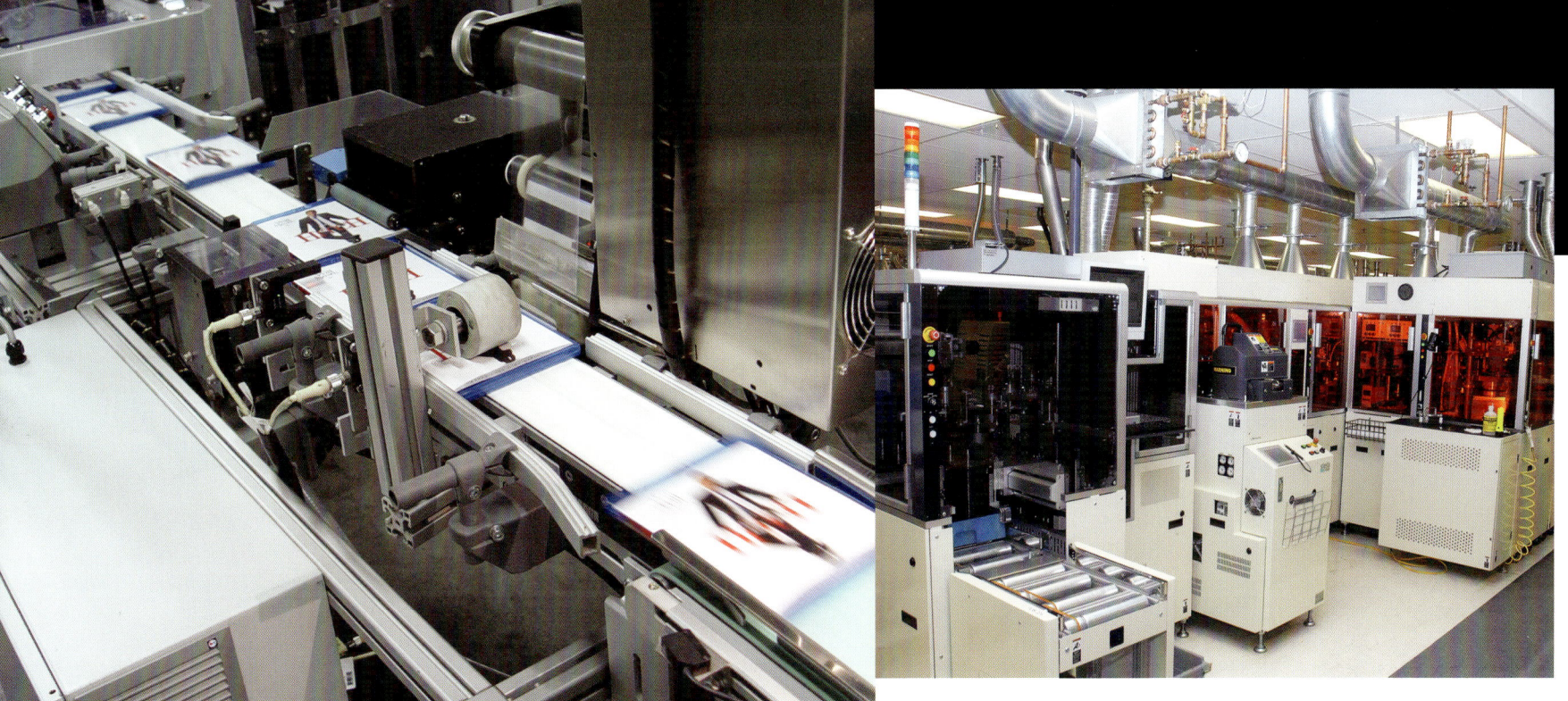

Both photos: © Hickman Photography

plant's total employment to upward of 1,300 employees, which represents a great boon to the local Indiana economy.

A Growing Supply Chain of Advanced Offerings

As Sony DADC continues to grow, it is also broadening its content-preparation services for audio, video, and multimedia content holders. These services—provided by the Sony DADC group called DigitalWorks—include everything from project planning and graphic interface design for all of Sony DADC's disc formats to non-disc-related services such as encoding, storage, and management of digital content directly to Internet Service Providers (ISPs) or other locations.

Two additional companies—Sony Entertainment Distribution and Corporate Freight Management—merged under the Sony DADC moniker in 2008. The addition of the services provided by these two companies allows Sony DADC to provide a more complete supply chain, featuring progressive customer service and increased control over timing and cost.

Sony DADC is both an ISO-9000 and ISO-14001 certified company and has also won additional recognition for operational excellence—proudly, it received the first-ever Indiana Quality Assurance Award in 1999 and won the coveted U.S. Senate Productivity Award in 2004. For more than a decade Sony DADC has not only adhered to environmental initiatives but it has increasingly gone beyond them. The facility recycles 98 percent of all waste produced, reduces CO_2 emissions via more than 70 internal Six Sigma projects, and encourages its employees to reduce their personal footprint with an employee recycling program and other efforts.

In 2009 Sony DADC celebrates 25 years of continuous operation. This company has grown into a valued employment provider for its local community and a globally respected provider of quality optical discs and technology services to its clients. More information about the company is available on its Web site at www.sonydadc.com.

Far left: The movie *Hitch* by Sony Pictures Home Entertainment was the first Blu-ray Disc™ (BD) to roll off the packaging line in May 2006 at the Terre Haute facility. Left: Twenty-six BD replication lines were added in the summer of 2008 to nearly double the capacity of the format to meet growing customer demand. A clean-room environment is required during the mastering of BDs to ensure that no dust or other particles mar the quality or functionality of the discs. Above: BDs are available in both a single- and dual-layered format. Single-layer BDs hold up to 25 gigabytes (GBs) of information and dual-layered BDs hold up to 50 GBs.

89

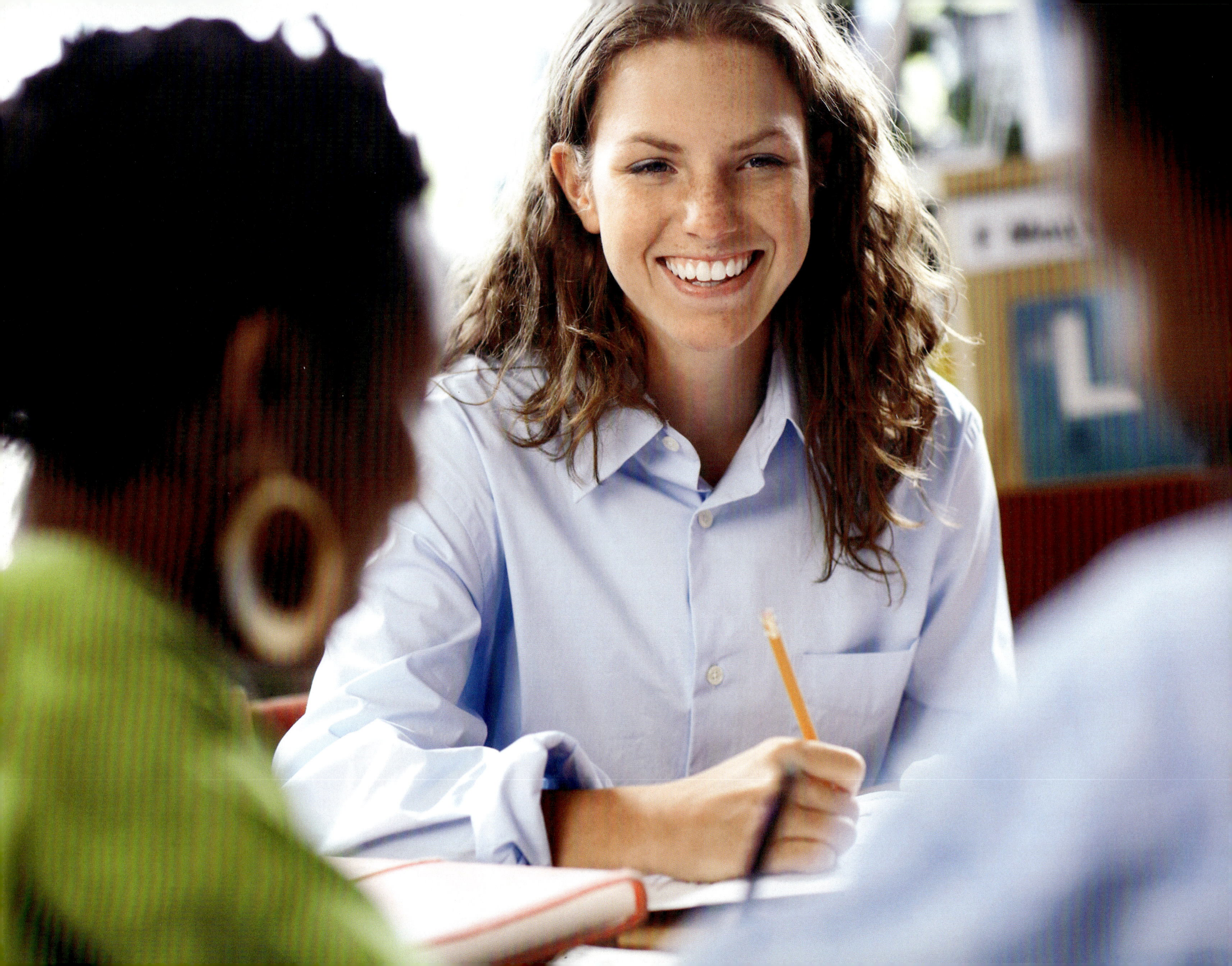

PROFILES OF COMPANIES AND ORGANIZATIONS
Education

Vincennes University

Committed to enhancing Indiana's educational opportunities and economic development, this institution offers quality instruction that is both affordable and accessible. With multiple sites and a wide range of majors, this school offers programs that lead to associate degrees, certificates, and select baccalaureate degrees.

Above: Vincennes University attracts students from 28 states and 34 countries. Students can enroll in baccalaureate, associate degree, and certificate programs.

Vincennes University (VU) serves an important and unique role in Indiana's postsecondary educational system. As the state's premier transfer institution, VU offers innovative career programming; developmental education; and proven associate, certificate, and baccalaureate degree programs. By customizing its curriculum to address the specific needs of Hoosiers in the classroom and in the workplace, VU has become a valuable asset to Indiana's workforce and economic development.

VU is Indiana's first college, established in 1801 by Indiana Governor William Henry Harrison, who went on to become the ninth U.S. President. In addition to the original Vincennes campus—which today spans 160 acres—the college has extended campuses in Jasper and Indianapolis.

Drawing on more than 200 years of history, VU has created a diverse, student-centered community. VU delivers the complete college experience through more than 50 clubs and organizations; annual events like homecoming and the Wabash River Tube Race; intercollegiate and intramural competitions; and a wide assortment of other cultural, educational, and recreational activities. Annually, VU serves nearly 11,000 students from Indiana, across the nation, and more than 34 countries.

VU offers the lowest tuition rates of any public, residential college in the state, providing a quality education at an exceptional value. With a low student-to-faculty ratio of 15:1, VU offers students a solid educational foundation from faculty members dedicated to teaching, not research. In addition to offering 200 associate degree and certificate programs, VU also offers bachelor's degree programs in technology, health care management, homeland security, nursing, and education.

An Educational Asset

VU is an important link in Indiana's educational system. Through its academic transfer programs, early-college opportunities, developmental education programs, and baccalaureate degree programs, VU educates students from all academic levels with a variety of goals.

VU is responsible for more associate transfer degrees than any other college in Indiana. Courses are designed to be compatible with the state's other higher education institutions, allowing VU graduates to transfer with confidence. Working with faculty at sister institutions, VU has developed high quality transfer options that meet the needs of students and contribute to Indiana's education opportunities.

Photo: © Vincennes University

Workforce Development

VU is dedicated to the development and enhancement of education and experience-based knowledge in Indiana to foster the state's economic growth and vitality. The university's technical training and workforce development programs help create a highly trained technical workforce of new and incumbent workers with up-to-date skills.

In the 1950s, VU responded to the need for a curriculum that included innovative career and technical education programs that would lead to immediate job placement after graduation. Ongoing program refinement and upgrades produce successful graduates that are in high demand. Nearly 95 percent remain in Indiana and contribute to the state's economic health.

VU's new State Center for Applied Technology designs programming customized to meet the training needs of the state's business and industry. Workforce development opportunities are available both on and off-campus using current technology. Additional advanced manufacturing centers in Dubois and Gibson counties will support future industry growth by delivering education and training programs for high-demand manufacturing, logistics, and technology fields.

Vincennes University's combination of quality instruction and innovative programming develops highly skilled and well-educated workers who help draw business and industry to Indiana. Through its dedication to adding value to higher education in Indiana, VU is helping Indiana build a stronger, more prosperous future. Additional information about Vincennes University is available at www.vinu.edu.

VU's dual enrollment and early-college partnerships allow students to add value to their high school education by simultaneously earning college credit. These programs help increase high school graduation and postsecondary entrance rates, improve the accessibility of postsecondary education, and ease the transition from high school to college.

For more than 30 years, VU's open-door admissions policy has included a developmental education option, paving the way for students of all ages and abilities to achieve academic success.

Since 2004 VU has offered a number of select baccalaureate degree programs that address shortages in Indiana's workforce.

Top left: Students enrolled in Audio Recording at VU use the computer-based Pro Tools system and may select a concentration of multimedia applications, history of popular music styles, or music business courses. The program emphasizes hands-on experience using industry-standard recording media. Below left: VU's Red Skelton Performing Arts Center features an 800-seat proscenium theater complemented by modern classrooms and laboratories. Top right: One of the nation's leading programs of its kind, VU's Advanced Manufacturing program produces graduates who are in high demand by many leading companies.

Photos: © Vincennes University

kWh

PROFILES OF COMPANIES AND ORGANIZATIONS
Energy, Fuel, and Utilities

BP Products North America, Inc.

This pioneering company, part of energy giant BP, plc, is making significant investments in its Indiana refinery. By reconfiguring its processing capacity to refine additional amounts of heavy crude, it is increasing economic and environmental benefits for the local community and diverse, secure energy sources for the nation.

Above: BP Products North America, Inc., owns and operates the iconic Whiting Refinery, which is located in northwest Indiana on the south shore of Lake Michigan.

BP Products North America, Inc. (BP Products), a wholly owned subsidiary of BP, plc, is involved in the exploration, development, production, refining, and marketing of oil and natural gas. It owns and operates five refineries in the United States, in Texas, California, Washington state, Ohio, and Indiana. BP Products' Whiting Refinery in northwest Indiana is the company's second-largest refinery, providing the nation with gasoline, diesel, jet fuel, and other products fundamental to society today.

A Pioneering Refinery

Whiting Refinery is credited with developing many of today's modern refining processes, and it has a long history of contributing to Indiana's energy demands. A landmark on the shore of Lake Michigan for more than a century, it began operating in 1890, and its first shipment consisted of 125 tank cars of kerosene. At the time, gasoline was considered to be a waste product. The initial production capacity of the refinery, then Standard Oil of Indiana, was 17,000 barrels of crude oil per day. It later became Amoco, and then BP. Whiting Refinery today has the capacity to process more than 400,000 barrels per day—making it one of the four largest refineries in the nation—and it continues to increase its capacity to produce additional supplies of motor fuel.

An important supplier of petroleum products for the Midwest, Whiting Refinery makes enough products each day to fuel 430,000 cars, more than 10,000 farm tractors, 22,000 semitrucks, and 2,000 commercial jets and to fill 350,000 propane cylinders. Additionally, it makes 8 percent of the asphalt produced in the United States. The refinery is located on 1,400 acres across Whiting, Hammond, and East Chicago, Indiana, and is operated around the clock, every day of the year. It employs 1,700 BP workers and from 500 to 1,500 contract workers.

Whiting Refinery has always been dedicated to developing innovative, efficient technology. By 1913 the facility's Drs. William M. Burton and Robert E. Humphreys had developed thermal cracking, which uses heat and pressure to increase gasoline yields. The refinery continued to develop important advancements in the petroleum industry, including its 1969 initiation of a valuable third step in the water-treatment process to remove impurities. In this tertiary stage, air is injected into the water, bringing impurities to the surface to be removed. The Whiting Refinery uses about 100 million gallons of water per day from Lake Michigan for cooling purposes. Lake water utilized in the refining process (as well as storm water) is treated at the refinery's wastewater treatment plant in strict compliance with all state and federal regulations before it is returned to Lake Michigan.

The refinery remains committed to using the latest technology and safety and environmental processes. Today BP Products is working on its Whiting Refinery Modernization Project. The project represents an investment of $3.8 billion, more than $1.4 billion of which is allocated for environmental improvements. Completion of the project is planned for 2011.

Improving Production and the Environment

This vast upgrade and modernization project will increase Whiting Refinery's daily motor fuel production by 1.7 million gallons as well as increase the facility's ability to refine heavy crude oil from Canada's Alberta province. Between 20 and 25 percent of the refinery's current feedstock is heavy crude oil. Once the upgrade is complete, this figure will increase to between 80 and 90 percent.

Modernization of the Whiting Refinery—which supplies 16 percent of the gasoline and 20 percent of the diesel and jet fuel used in the Midwest—will include constructing a new coker, a crude distillation unit, a gas oil hydrotreater, and sulfur-recovery facilities, as well as upgrading the refinery's water-treating facilities. The project will benefit the northwest Indiana community in a number of ways. The refinery will continue to make important economic contributions to the region. Direct local spending during construction, including salaries and wages, will be in excess of $2.5 billion. Up to 2,000 craftsmen will be employed by 2010. The project provides a significant increase to the tax base of northwest Indiana. Another important aspect of the project is its $1.4 billion budget for the design and integration of improvements that protect the environment. The refinery's upgrades represent the largest private corporate investment made in Indiana and demonstrate the company's commitment to the region.

Canadian Crude Project

By upgrading Whiting Refinery to be able to process a greater amount of heavy crude oil, BP Products will be increasing the reliability of the United States' fuel supply, creating greater energy security for the entire country. In addition, Whiting's potential use of Canada's heavy oil resources in the future could mean the refinery could purchase up to 80 percent of its crude oil at significantly lower costs.

Most of Canada's oil sands or about 178 billion barrels of recoverable reserves are held in a 23,000-square-mile area of northeastern Alberta.

The oil sands contain bitumen, which is denser than conventional oil and has more impurities. It is therefore harder to refine and requires that important refinery upgrades be made before it can be processed. Canada is the largest energy supplier to the United States, and the production of the Canadian oil sands is anticipated to more than double by 2013. BP Products is investing in its Whiting Refinery so that it will be able to process Canada's vast, reliable source of heavy crude oil. With declining reserves of lighter crude oil in the United States, BP Products' plans for using Canada's heavy crude oil will help the United States to meet its need for diverse, secure energy sources.

By improving the United States' energy security and bringing thousands of jobs to the region through the Whiting Refinery Modernization Project and the Canadian Crude Project, BP Products North America will continue to be an important economic asset to the northwestern Indiana community. Additional information is available on the company's Web site at www.bp.com.

Above left: BP Products is in the process of upgrading and modernizing its Whiting Refinery. The project benefits the nation's energy security and fuel supply reliability and sustains the refinery's positive contribution to the region's economy, providing jobs for thousands of BP employees and contractors in northwest Indiana.

ProLiance Energy

With a reputation for being a secure and reliable energy provider, this natural gas marketing and trading company provides customers with a cost-effective supply of natural gas and risk-management products while offering the community its time, talent, and financial support.

Above: ProLiance Energy provides a wide range of customers with resource management and a reliable, cost-effective natural gas supply to meet heating and manufacturing process needs.

Headquartered in Indianapolis, ProLiance Energy is a full-service natural gas marketing and trading company with four sales and service offices throughout the Midwest. Founded in 1996, the company has seen its customer base grow to over 2,400 locations, including education, health care, commercial, and industrial facilities as well as municipalities, power generators, and utilities. These organizations and many others know they can trust ProLiance Energy to best serve their natural gas needs.

Committed to a high standard of excellence and complete customer satisfaction, ProLiance Energy is proud to have earned top rankings in an industrywide study conducted for natural gas marketers in 2006 and 2007 by Mastio & Company, an international research and consulting firm that focuses on commodity-driven industries. The study identifies the factors that determine how natural gas buyers choose their suppliers. ProLiance Energy was among the top eight companies that exceeded the industry benchmark.

Products and Services

ProLiance Energy has become an industry leader by meeting customers' needs, even during periods of volatility. This is achieved by effectively utilizing physical and pipeline storage, conducting thousands of transactions per year on more than 15 pipelines, and moving an average of one billion cubic feet of natural gas per day. Experienced account managers focus on identifying customers' needs, setting goals, and exceeding expectations. ProLiance Energy provides expertise, effective solutions, and educational information so that customers can make informed natural gas purchasing decisions.

At ProLiance Energy, the goal is to provide the highest level of customer service and deliver a reliable supply of natural gas at all times. Reliability, combined with timely pricing and forecasting information, prompt contract negotiations and execution, accurate and timely billing, flexible contract terms and pricing, and innovative risk management products, assures customers that a relationship with ProLiance Energy is a true partnership.

With a continuous focus on service, ProLiance Energy has developed a full spectrum of Signature products, which are designed to assist in risk management, monitoring and measuring energy usage, and creating energy savings and sustainability solutions. These products help customers realize cost savings and efficiency gains, in addition to strengthening corporate environmental stewardship.

Commitment to the Community

Innovation, service, trust, and reliability filter through everything ProLiance Energy does. The company takes responsibility for serving the community as well as serving customers. Whether demonstrated through financial support, time, or talent, ProLiance Energy upholds a commitment to the civic, cultural, and athletic endeavors that contribute to the communities in which its employees live and work.

From the top down, ProLiance Energy employees wholeheartedly support their community. United Way of Central Indiana is just one organization that benefits from the company's participation; in 2008 ProLiance Energy was named a United Way of Central Indiana Spotlight company. One hundred percent of ProLiance employees contribute to United Way. Numerous employees participate in the BackPack Attack, a school-supply drive for students in need, and many participate in ReadUP, a tutoring program designed to help improve reading skills in public schools.

In addition to partnering with United Way, ProLiance Energy employees are active in organizations such as the Cystic Fibrosis Foundation, Big Brothers Big Sisters, Indiana Repertory Theatre, the Indianapolis Parks Foundation, and many other worthy causes. The company is proud to support more than 60 cultural, educational, health care, and youth organizations throughout Indiana.

Partnerships

ProLiance Energy is proud to partner with customers and key organizations within the community. Partnerships have been developed with several organizations, such as the Indiana Economic Development Corporation, the Indiana Manufacturers Association, the Indiana Chamber of Commerce, the Indiana Association of School Business Officials, and the Indiana Society of Healthcare Engineering. These partnerships enhance ProLiance Energy's ability to serve customers and communities.

Additional information about ProLiance Energy is available on its Web site at www.ProLiance.com.

Far left: ProLiance Energy supports the "Keep the Heat On" campaign to offset the cost of heating facilities that are operated by not-for-profit organizations. Left: ProLiance Energy procures and manages more than 300 billion cubic feet of natural gas for its customers. The company's staff of more than 100 seasoned industry veterans includes a highly qualified sales and trading staff, which provides competitive, reliable natural gas supplies for price-sensitive customers; and a supply and operations staff, which has extensive experience at managing natural gas deliveries throughout the United States. The entire staff's working knowledge of the energy industry provides the foundation for ProLiance's customer-focused service and reliable supply.

Duke Energy Corporation

This Indiana electric utility, one of the nation's largest, serves more than 780,000 customers directly and provides wholesale power to rural electric cooperatives and municipalities. Its mission is to make people's lives better by providing energy services in a way that balances environmental, economic, and social priorities.

Electricity "rode in by rail" in Indiana. Electricity began to flourish in the state in the late 1800s when electrified city street railways became available. The same generators that supplied power to run streetcars had excess capacity that companies sold to anyone wanting it. The street railways inspired the idea of operating electric rail cars, called interurbans, between cities. The interurbans accelerated the development of an electric power network across the state.

Public Service Indiana, a major league utility, evolved from the interurban companies. The company became the state's largest electric supplier and eventually would change its name to PSI Energy. In 1994 PSI Energy merged with Cincinnati Gas and Electric Co. to form Cinergy. And in 2006 Cinergy merged with Duke Energy Corporation, which supplies and delivers electricity to approximately four million customers in Indiana, Ohio, Kentucky, and the Carolinas. The company has approximately 35,000 megawatts of electric generating capacity in the Midwest and the Carolinas and natural gas distribution services in Ohio and Kentucky. Duke Energy trades on the New York Stock Exchange under the symbol DUK.

With some of the lowest electric rates in the nation, Duke Energy's Indiana operations have helped fuel the state's economic development. While its network of primarily coal-fired power plants has delivered low rates, they also come with an environmental responsibility. The company has invested more than $1 billion in pollution control at its Indiana plants to lessen their environmental impact. The company also has been a leading voice for public policies on the environment. It broke with the industry to support the 1990 Clean Air Act amendments and, more recently, advocated for action on greenhouse gases, a driver of climate change.

Throughout its evolution, the company also has been a force in its communities. Financially, each year Duke Energy dedicates a portion of its pretax earnings from its regulated operations to not-for-profit organizations in the states it serves. Equally important is that it couples charitable giving with time off and support for employee volunteerism.

From the days of the interurban railways until now, the company's role as a provider of power that fuels Indiana makes it a central player in the state's economy, history, and community life. Additional information is available on the company's Web site at www.duke-energy.com.

Right: In Indiana, Duke Energy Corporation serves customers in portions of 69 counties through more than 25,000 miles of transmission and distribution lines.

Photo: © Tom Wolfe

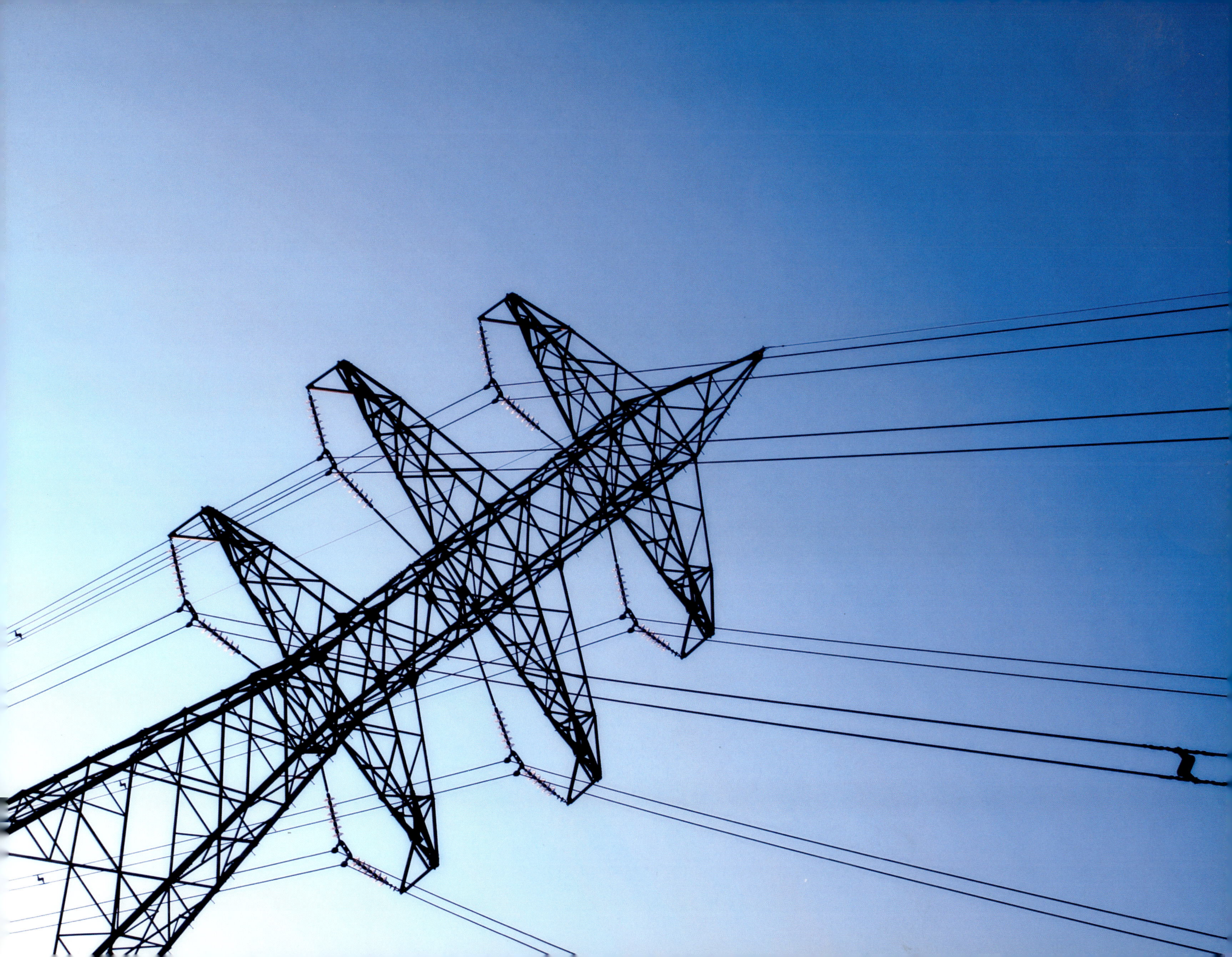

A Division of NiSource Inc.

Headquartered in Merrillville, this company provides natural gas transmission, storage, and distribution as well as electric generation, transmission, and distribution to northern Indiana. The company is continually exploring the best options to meet the needs of its residential, commercial, and industrial customers.

Right: Located in White County near the town of Monticello, Indiana, the Norway Hydroelectric Dam is one of two hydroelectric facilities along the Tippecanoe River owned and operated by Northern Indiana Public Service Company (NIPSCO), a division of NiSource Inc. Below right: NIPSCO employs more than 2,500 workers who are committed to providing safe and reliable natural gas and electric service to the company's customers.

Northern Indiana Public Service Company (NIPSCO)—which can trace its origins back to 1853—was established under its present corporate form in 1926. Samuel Insull was chosen as the company's first chairman.

Today NIPSCO is Indiana's largest natural-gas distribution company and the second-largest electric distribution company in the state, providing safe and reliable energy to more than 712,000 natural gas customers and 445,000 electric customers across the northern third of Indiana.

NIPSCO is owned by NiSource Inc. (NYSE: NI), a utility holding company that operates 10 regulated utilities in nine states and two major interstate pipelines from the Gulf of Mexico to New England. Both companies share a headquarters in Merrillville, Indiana. NiSource is one of only four Fortune 500 companies headquartered in Indiana and one of 15 companies that make up the Dow Jones Utilities Index.

With more than 2,500 employees, NIPSCO operates three coal-fired plants and has a renewable-energy portfolio that includes wind power and two hydroelectric-generating sites, while maintaining more than 15,000 miles of natural gas pipeline across Indiana.

Additionally, NIPSCO owns and operates a 535-megawatt (MW), combined-cycle natural gas facility. Since concern is increasing over carbon emissions, natural gas–fired plants are considered one of the best environmental options for power generation.

In a continued effort to support the growth of Indiana's economy, NIPSCO makes significant investments in state, regional, and local economic development organizations. The company is proud to partner with these groups and hundreds of other community leaders to attract new businesses to the area, encourage new investment by existing businesses, and help create new, higher-paying jobs. NIPSCO also supports more than 600 local organizations across northern Indiana through charitable giving, partnerships, and volunteer work. Additional information about NIPSCO can be found on the company's Web site at www.nipsco.com.

By taking an active part in supporting quality growth and development across northern Indiana, NIPSCO remains committed to providing its customers with safe and reliable service, its investors with a fair rate of return, its employees with careers that allow them to enjoy a good standard of living, and its communities with the chance to prosper.

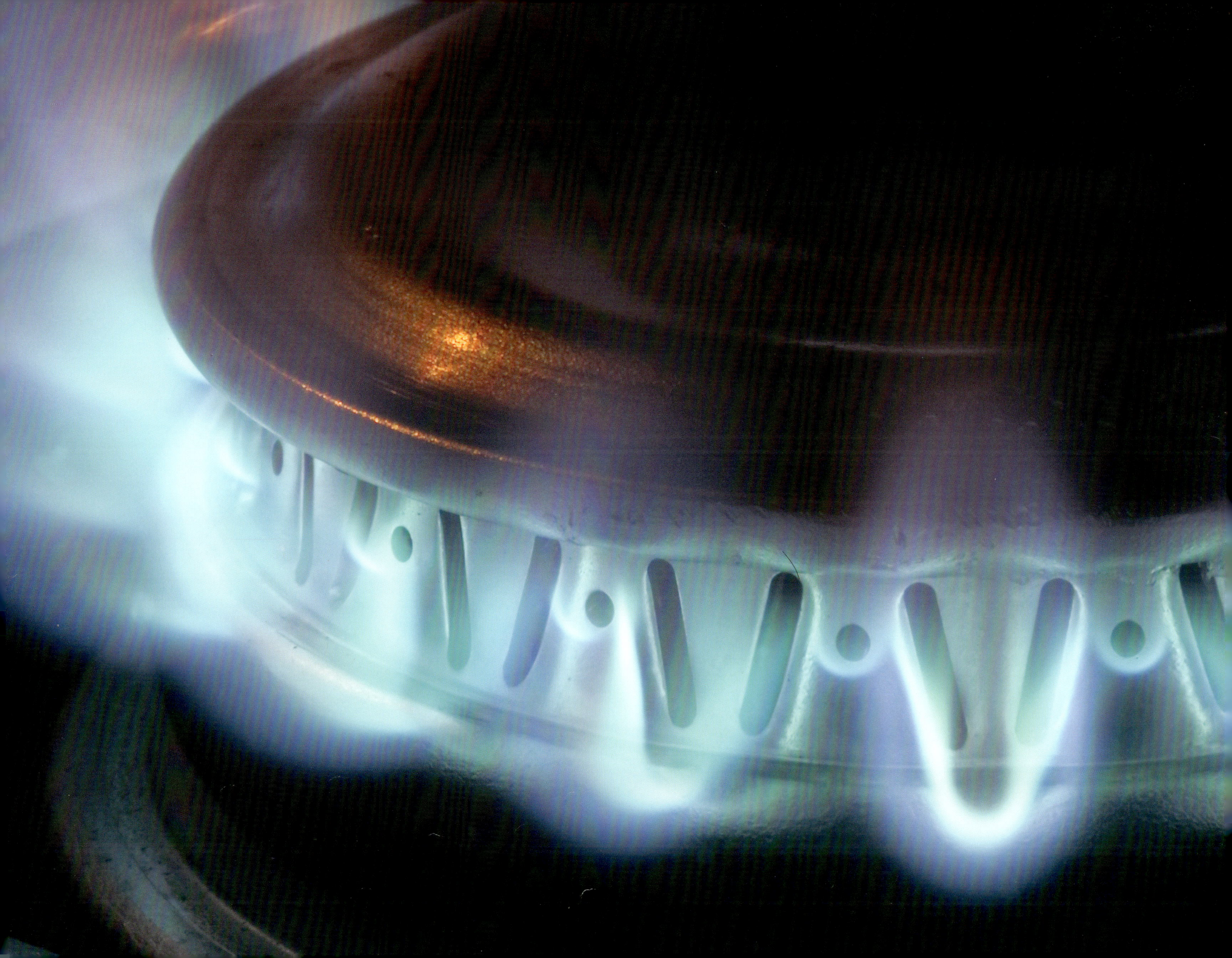

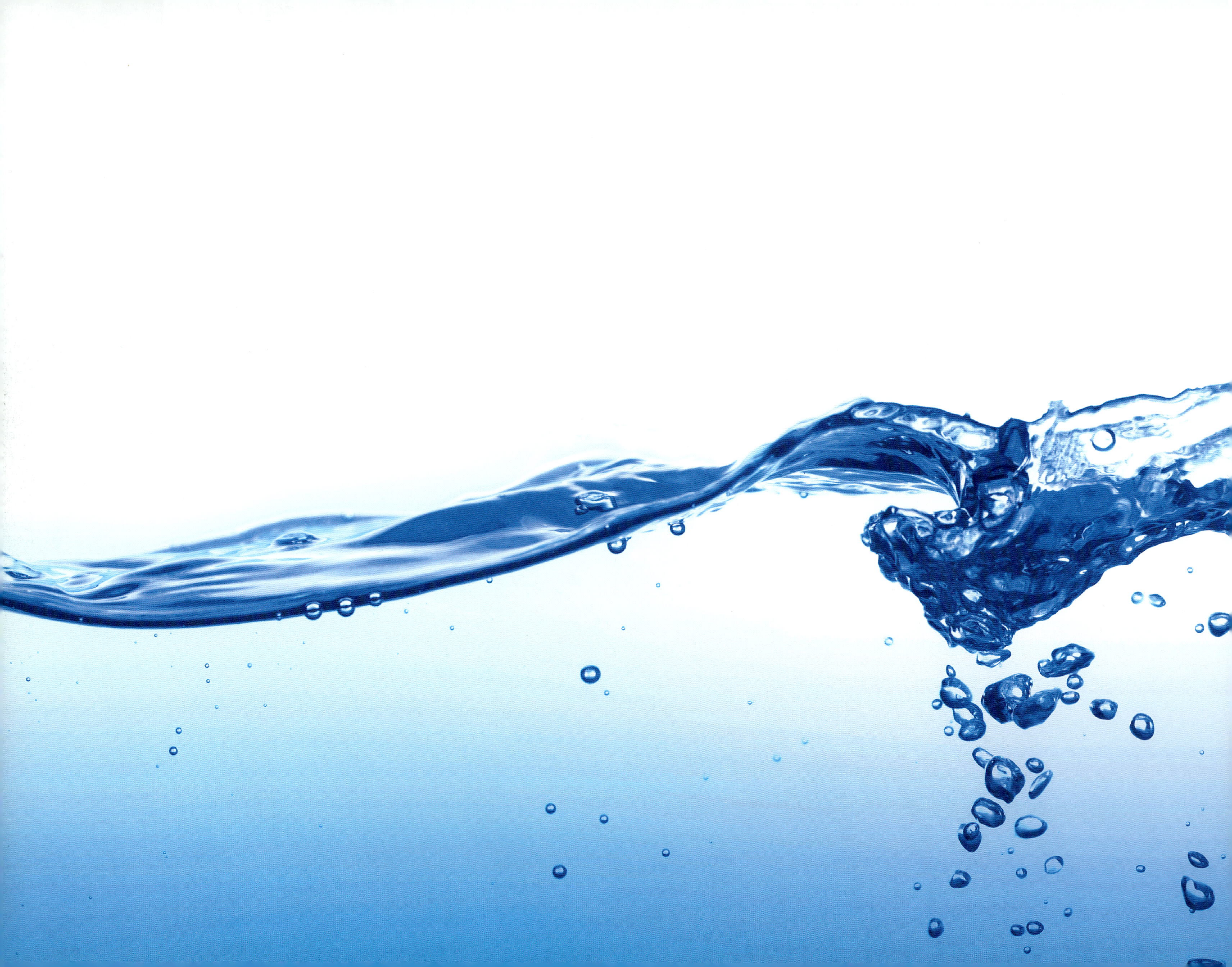

PROFILES OF COMPANIES AND ORGANIZATIONS
Environmental Resources

Indiana American Water

Dedicated to providing high quality water service to 126 communities throughout Indiana, this water utility company leads the industry in quality standards, continually invests in its advanced infrastructure, responsibly cares for the environment, and supports its communities.

Above: Indiana American Water is committed to delivering high quality water service to all of its customers in Indiana. Right: A water sample is taken for testing at one of the company's wells along the banks of the Ohio River.

Indiana American Water, a subsidiary of American Water (NYSE: AWK), takes pride in its role as a steward of water, a vital element of life. This innovative utility company annually pumps, treats, and delivers more than four billion gallons of high quality water to more than 1.2 million customers in Indiana. It utilizes a variety of surface water and groundwater sources to serve its customers. At all of its 32 water treatment plants across the state, the processes are tailored to local water resources and community needs.

It strictly follows the regulations set out by the Indiana Utility Regulatory Commission and the Indiana Department of Environmental Management to ensure a fair price for excellent quality water and wastewater service. All four of its eligible local surface water treatment operations have been distinguished five times with the United States Environmental Protection Agency (USEPA) Director's Award—an achievement accomplished by only 100 water suppliers—which recognizes water suppliers whose standards surpass USEPA requirements. Annually it conducts over 34,000 tests for about 100 contaminants and checks the quality of its drinking water throughout the water treatment and delivery process.

Because American Water works in cooperation with the USEPA to help in the development of national drinking water standards, it is often years ahead of the industry in implementing regulations. American Water is also a proud partner in the USEPA's WaterSense® Program. Launched in 2006 to encourage water efficiency and water saving practices in homes, the program brings together almost 1,000 partners across the nation to raise awareness of sustainable water solutions ranging from high-efficiency toilets to certification programs for irrigation professionals.

Indiana American Water is committed to ongoing investment in its infrastructure so it can deliver excellent water and wastewater service at an exceptional price. Water is still a great utility value at a cost of less than a penny a gallon. Since 2000 the company has invested more than $310 million in system improvements such as extending water mains, replacing meters and hydrants, upgrading plant equipment, and constructing new treatment facilities. Recent major projects include a new, expandable water treatment facility in Kokomo; a new, one-million-gallon elevated water storage tank in Portage; a $7.6 million system expansion project in Noblesville; and three new groundwater treatment plants to serve customers in West Lafayette, Greenwood, and Shelbyville.

As a water and wastewater service provider, environmental stewardship is central to Indiana American Water's mission. The company demonstrates its commitment with ongoing projects through the state that protect water resources, preserve wildlife habitats, and help threatened species. Indiana American Water is dedicated to maintaining water supply for future generations through a variety of programs in each of the communities it serves.

A good neighbor, Indiana American Water invests time and financial resources to give back to its partner communities. Among the programs it supports are local United Way campaigns and Water for People, a program that helps improve the quality of life for impoverished people around the world. It also sponsors the "Pathway to Water Quality" exhibit at the annual Indiana State Fair, participates in Earth Day celebrations, and provides an environmental grant program to encourage community environmental projects.

Committed to delivering high quality water and wastewater service to its customers, Indiana American Water will continue to expertly care for Indiana's water every day. Additional information about Indiana American Water can be found on the company's Web site at www.indianaamwater.com.

Left: Indiana American Water is dedicated to investing in its infrastructure to meet the needs of the nearly one in five Indiana residents that depend on the company for quality water and wastewater service each day.

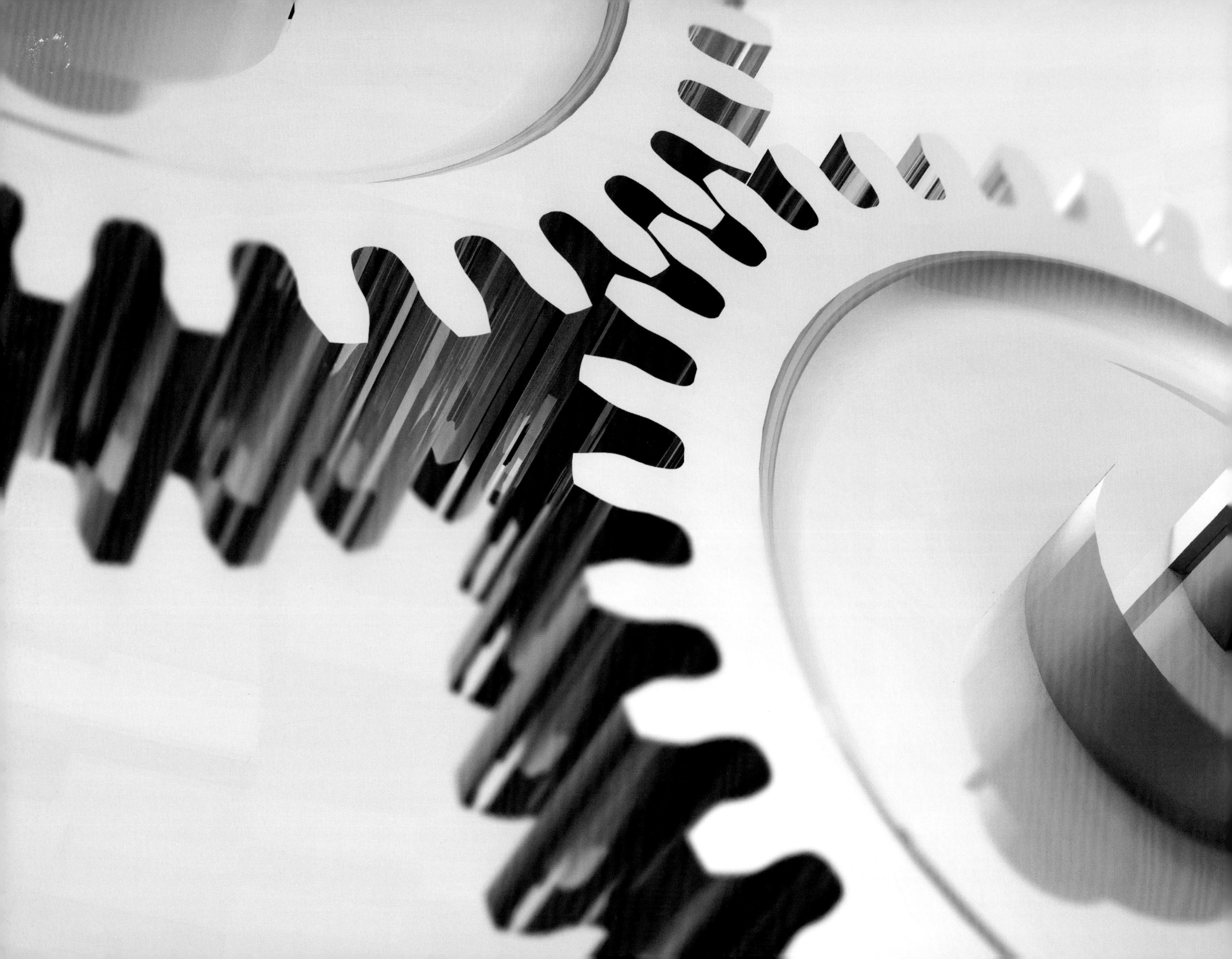

PROFILES OF COMPANIES AND ORGANIZATIONS
Manufacturing and Distribution

Steel Dynamics, Inc.

In just 15 years this homegrown Hoosier company has branched out into several states and become one of the most profitable steel producers in America. It is the nation's fifth-largest producer of carbon-steel products with 2006 annual revenues of $3.2 billion and annual shipments of 4.7 million tons.

Above: Sparks fly at the Steel Dynamics Flat Roll Division melt shop in Butler as the electric-arc furnace is tapped and emptied into ladles below. Right: Finished beams are loaded onto a rail car at the Structural and Rail Division.

In 1993 steel-industry veterans Keith Busse, Mark Millett, and Dick Teets set out to do something that had not been done in the United States for many years—create a new, independently financed American steel company. The new venture would be called Steel Dynamics, Inc. (SDI).

Conventional wisdom had it that American steel was doomed, that domestic steel production would continue to decline until American industry would become largely reliant on foreign steel suppliers. In the information age, some thought, there would be little room for smokestack industries, much less a new steel manufacturer.

Busse and company thought differently. They were certain that their talented, resourceful engineers—using new production techniques—could build mills that when operated by motivated, talented employees would compete effectively with any foreign competitor in terms of both cost and quality. They also believed that it was clearly in the United States' best interest to maintain a strong, vibrant domestic steel industry.

Fortunately, a group of investors shared their belief, allowing SDI to construct a world-class minimill in northeastern Indiana. Built in a record 14 months, the flat-roll mill in Butler, Indiana, now sets the pace in productivity and profit margins in world production of hot-rolled coils.

Following the first shipments of hot-band steel from Butler early in 1996, SDI grew rapidly. It is now the nation's fifth-largest producer of carbon-steel products and is among the most profitable American steel companies in terms of profit margins and operating profit per ton.

In 2000 SDI invested in New Millennium Building Systems, which uses SDI steel to fabricate joists and other building components for commercial buildings.

SDI's second greenfield minimill, located at Columbia City, Indiana, began production of structural steel in the summer of 2002. The Structural and Rail Division now ships a wide variety of wide-flange beams. It made

All photos, both pages: © Steel Dynamics, Inc.

Steel Dynamics, Inc. Facilities

Steel Dynamics, Inc. operates five steel mills and finishing, fabrication, metals-recycling, and iron-unit-production facilities in 15 states and in Canada.

- Flat Roll Division, Butler, Indiana
  - Flat-Roll Finishing Facility, Jeffersonville, Indiana
  - The Techs, Pittsburgh, Pennsylvania
- Structural and Rail Division, Columbia City, Indiana
- Engineered Bar Products Division, Pittsboro, Indiana
- Roanoke Bar Division, Roanoke, Virginia
- Steel of West Virginia, Huntington, West Virginia
- New Millennium Building Systems, with locations in Indiana, Florida, Ohio, South Carolina, and Virginia
- OmniSource, Fort Wayne, Indiana, with locations throughout the eastern United States and Canada
- Iron Dynamics, Butler, Indiana
- Mesabi Nugget, Hoyt Lakes, Minnesota

its first rail shipments for railway use in 2004. The recent start-up of a brand-new, medium-section rolling mill and increased casting capability will enable the Columbia City mill to rival the Butler flat-roll mill's output.

SDI acquired and refitted an idled minimill in Pittsboro, Indiana, in 2002, forming the Engineered Bar Products Division, which makes and finishes special-bar-quality (SBQ) products to exacting customer standards.

In April 2006 SDI bought the Roanoke Electric Steel Corporation in Virginia, which was renamed the Roanoke Bar Division. This allowed SDI to expand its product offerings in merchant bars and structural shapes as well as to add to the New Millennium fold three joist-fabricating operations in the eastern United States.

Meanwhile, SDI's state-of-the-art Butler mill remains a world leader in minimill production of flat-rolled steels, consistently breaking its own records for productivity. The Flat Roll Division's finishing facilities at Butler and at Jeffersonville, Indiana, produce pickled, cold-rolled, galvanized, and Galvalume® sheet and painted, flat-roll steel. Also complementing the division's product mix was the 2007 acquisition of The Techs, which operates three modern flat-roll galvanizing plants in Pittsburgh.

SDI is now involved in backward integration as a hedge against the rising cost of scrap, the main ingredient in its steel. At Butler, SDI-owned Iron Dynamics produces lower-cost, iron-rich units as feed stock for the mill's electric-arc furnaces. Steel Dynamics is also initiating iron mining and iron-nugget production in Minnesota. And SDI has acquired Fort Wayne–based OmniSource Corporation, one of the nation's largest metals recyclers and the main supplier of scrap to its Indiana mills.

Today SDI is among the most successful American steel companies. The company continues to achieve record production and revenues. SDI consistently attracts talented, highly motivated people who thrive in an incentive-rich environment where dedication and innovation are well rewarded. Hallmarks of the culture are teamwork and respect for every employee at every level. New ideas always find an audience and often lead to process innovations.

Most importantly, however, Steel Dynamics, Inc. continues to set records for safety, striving to ensure that its employees are able to enjoy the fruits of their labor with their families. The company provides more information on its Web site at www.steeldynamics.com.

Top left: Special-bar-quality (SBQ) bars—used in a variety of industrial machinery and transportation applications—await shipment from the finishing facility at the Engineered Bar Products Division in Pittsboro, Indiana.

Left: A giant ladle of molten steel moves to the ladle metallurgy furnace where alloys are added and the steel's temperature is stabilized in preparation for casting.

Old Hickory Furniture Company

For over 100 years, this company has created high quality furniture out of the sturdy hickory saplings that grow in southern Indiana with an emphasis on old-fashioned craftsmanship. Its product offerings include indoor and outdoor furniture, lighting, wall art, bedding products and curtains, and custom kitchen and bath cabinetry.

Above: Old Hickory Furniture Company products are handcrafted, making each piece unique. Right, both photos: Old Hickory has furnished many national park lodges, including Old Faithful Inn at Yellowstone National Park (top) and Grand Canyon Lodge at Grand Canyon National Park (below). Opposite page: Old Hickory makes an array of products, from custom-made kitchen and bath cabinetry to indoor and outdoor furniture to lighting fixtures and pillows, blankets, and curtains.

Old Hickory Furniture Company has a long history in Indiana. The early pioneers that settled in the state recognized the abundant hickory saplings, which were both durable and flexible, as a good source of wood for building furniture. One pioneer in particular, Billy Richardson of Martinsville, began building handcrafted hickory chairs in the 1880s and selling them from his horse and wagon. In 1892 he and a few partners established the Old Hickory Chair Company, which was incorporated in 1898. The young company shipped its high quality, handcrafted furniture across the country, setting the standard for excellence in its industry. Over its 100-year history, the company has shipped its products to thousands of homes and resorts around the world.

Renamed the Old Hickory Furniture Company in 1922 to emphasize its broad offerings, which included beds, case goods, and upholstery items, the company relocated to Shelbyville in 1982. Today it continues Richardson's tradition of sophisticated artisanship at its 100,000-square-foot factory and headquarters where 150 employees create durable, comfortable, and unique hickory furniture by hand. From the initial harvesting of hickory saplings and curing of the wood in a kiln to the handwoven upholstery and final inspection, each piece of furniture goes through from 10 to 15 stages before it leaves the factory.

Carefully measuring and matching each unique piece of wood before assembling the frame, the company's craftspersons are attentive to detail, creating sturdy furniture that lasts for decades. A testament to the durability of the company's products, the Old Faithful Inn at Yellowstone National Park still uses the Old Hickory chairs that it installed in its dining room in 1906. Old Hickory has been serving the hospitality industry since 1899.

To offer its customers products for the entire house, Old Hickory has expanded its line of products to include lighting fixtures, custom-made cabinetry, and soft goods, including blankets, pillows, and curtains. Through a partnership with renowned wildlife artist Rod Crossman, Old Hickory also sells his artwork, which is available inlaid on select furniture collections, printed on leather-backed chairs, or as signed and framed prints.

Combining these new products with its emphasis on quality and craftsmanship, Old Hickory Furniture Company continues to grow in popularity, creating rustic furniture that is both durable and comfortable.

Old Hickory Furniture Company provides additional information on its Web site at www.oldhickory.com.

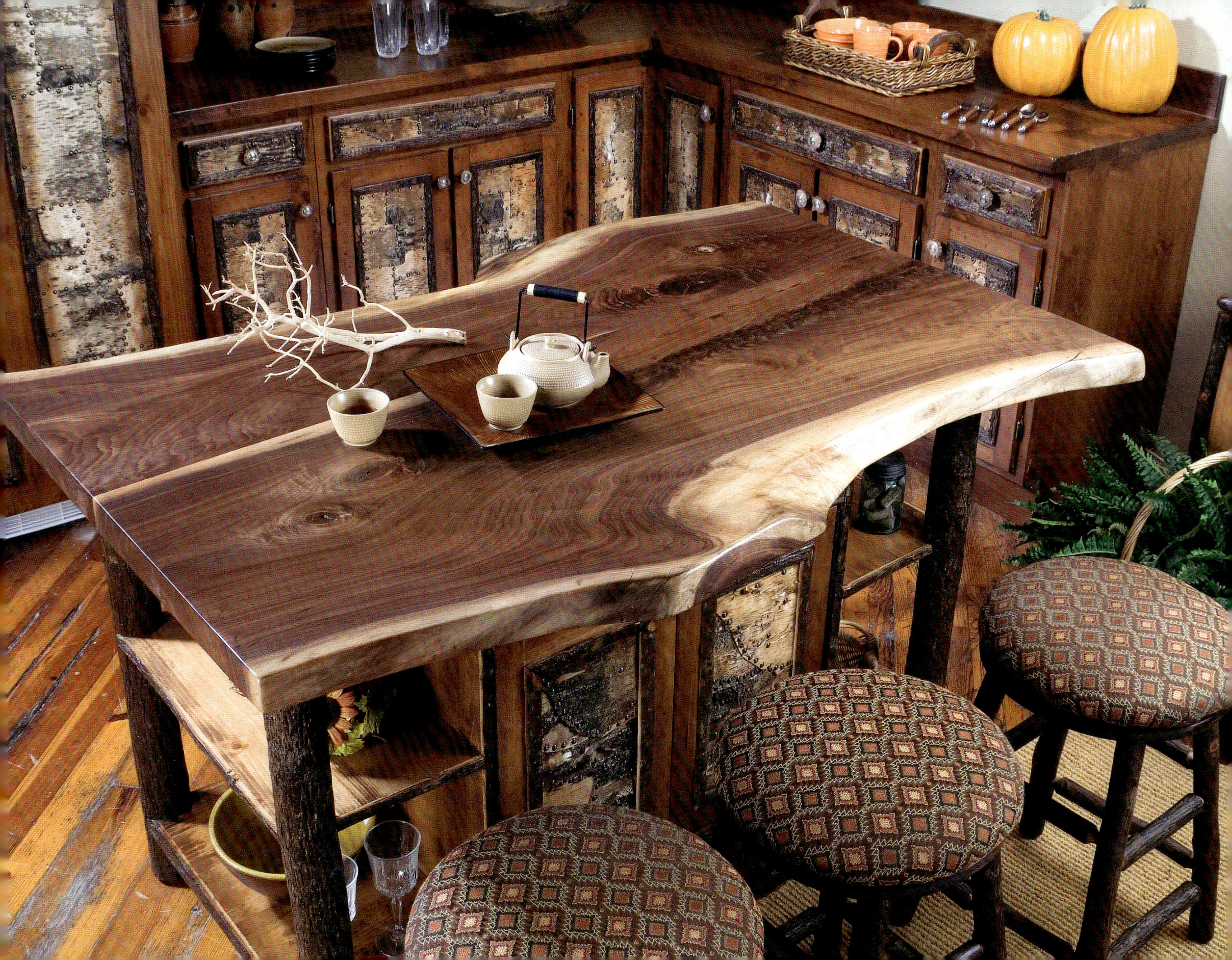

United States Steel Corporation

Guided by a new vision for its second century of business, this recognized leader in the global steel industry remains committed to making steel — its core focus for over 100 years — as well as strengthening its position in the global marketplace and building value for its employees, shareholders, and the communities in which it operates.

Above, all photos: United States Steel Corporation's (U. S. Steel's) Gary Works (left) is located on the south side of Lake Michigan. Composed of steelmaking and finishing facilities, Gary Works and its sister facilities — East Chicago Tin in East Chicago (center) and Midwest Plant in Portage (right) — are the company's largest combined manufacturing facility.

For more than 100 years, United States Steel Corporation (U. S. Steel) has provided steel products that have helped build, defend, and transport America. Many products that are a part of everyday life, such as automobiles, soup and beverage containers, washers and dryers, and various building materials, are made with high-quality, value-added products from U. S. Steel. The corporation traces its roots to 1901, when a group headed by two of America's most legendary businessmen, Judge Elbert Gary and John Pierpont Morgan, bought Andrew Carnegie's steel company and combined it with their holdings to create U. S. Steel. At the time, it was the largest business enterprise ever launched, and the company continued to grow.

Judge Gary, U. S. Steel's first chairman, understood the importance of effectively serving customers across the nation, so when he began looking for a place to build a new, state-of-the-art steel mill shortly after the company was formed, he knew his native Midwest would be the perfect location. A facility in this location could service the growing Western customer base as well as the steady demands of the East Coast market. Judge Gary chose the area now known as Gary, Indiana, a place with access to Lake Michigan and a vast rail network.

In early 1906, workers began excavating the sand dunes for what would become Gary Works. The mill was finished two years later, but the City of Gary sprang to life sooner than that with the completion of hundreds of homes, as well as churches, a hotel, a restaurant, a hospital, a library, and a YMCA. When production at Gary Works began in early 1909, it marked the start of a century-long relationship between an iconic American company and Northwest Indiana.

A Continuing Legacy

Today, U. S. Steel continues to have a significant presence in Northwest Indiana. Gary Works remains an industry leader and the company's largest plant. U. S. Steel also owns two steel finishing facilities in the area: East Chicago Tin in East Chicago and Midwest Plant in Portage.

Together, these three facilities provide family-sustaining jobs for more than 6,000 employees, as well as for people who work for numerous local companies that provide products or services to the plants.

Steel production begins at Gary Works, where coal is transformed into coke and combined with other raw materials in one of four blast furnaces to produce liquid iron. The iron is then mixed with scrap steel and different alloys in a basic oxygen process (BOP) shop to form the custom steel blends ordered by customers. The liquid steel is then cast into slabs that are moved to the hot strip mill, where they are rolled into thin sheets of steel and coiled.

Some coils then move on to the finishing facilities at Gary Works, East Chicago Tin, or Midwest Plant to undergo further rolling, treating, or coating before they are shipped to customers in the automotive, appliance, food packaging, and construction markets. With total raw steel production capability of 7.5 million net tons per year, Gary Works and its sister plants in East Chicago and Portage continue U. S. Steel's long-standing tradition of producing the high-quality steels needed by its customers while maintaining an intense focus on safety, environmental stewardship, product quality, and process efficiency.

A Commitment to the Community

One of U. S. Steel's objectives is to be a good corporate citizen in the communities in which it operates. To that end, the company is dedicated, first and foremost, to the safety of the men and women who work in its Northwest Indiana facilities. Safety is a core value at U. S. Steel. Training, employee engagement, and equipment investments are just a few of the things U. S. Steel is doing to eliminate all incidents and injuries and to help to ensure that its employees return home safely each and every day.

That same dedication also extends to environmental stewardship. Significant investments in advanced technology coupled with extensive employee training play a significant role in the company's efforts to protect Northwest Indiana's natural resources. Gary Works also created and continues to support a 20-acre oak savannah prairie habitat on its property next to the Indiana Dunes National Park, where an endangered butterfly species thrives along with native plants and animals.

Like the company, U. S. Steel employees care for and give back to the communities in which they live and work, and Northwest Indiana is no exception. Employees of Gary Works, East Chicago Tin, and Midwest Plant commit a considerable amount of their resources and free time to charitable and philanthropic endeavors that improve the quality of life in the region.

Having entered its second century, United States Steel Corporation remains committed to the same values that guided its first 100 years, and its facilities in Northwest Indiana — Gary Works, East Chicago Tin, and Midwest Plant — play important roles in the company's efforts to maintain its reputation as a world leader in the global steel industry.

Above, both photos: Gary Works remains a top producer in the industry thanks to its technologically advanced ironmaking, steelmaking, and finishing operations.

MasterBrand Cabinets, Inc.

One of the largest and most successful cabinetmakers and distributors in America, this legendary company offers a complete spectrum of products, from ready-to-assemble to custom-made, that are designed for the kitchen, bath, and office and address each customer's individual needs.

Above: MasterBrand Cabinets, Inc. designs and constructs custom kitchen, bath, and office cabinetry.

Opposite page: Long regarded as an industry leader, Aristokraft delivers on its promise of "Great Looks. Great Value."

MasterBrand Cabinets, Inc. makes more than cabinets—the company makes dreams come true. For customers who are improving their home or building their dream home, MasterBrand creates living spaces that meet their specific needs. Offering a variety of styles for every taste and budget, the company adheres to four basic principles that have made it a powerful force in the kitchen and bath industry: deliver complete; deliver on time; meet or exceed quality standards; and offer competitively priced, fashionable products.

Headquartered in Jasper, Indiana, MasterBrand makes and markets residential kitchen, bath, and home cabinetry under many well-known brand names. Aristokraft includes popular, affordable styles backed by the Good Housekeeping Seal of Approval. Decora is a semi-custom line with exceptional design flexibility. Schrock's Smart Solutions incorporates many organizational features that offer flexibility, accessibility, and versatility. And Diamond Logix cabinets, recommended by the Arthritis Foundation, are designed to make objects easier to reach and grasp.

One of the nation's and the world's largest cabinetry manufacturers, MasterBrand has 30 locations in the United States and overseas, employs 12,000 workers, and supplies multiple channels of cabinet distribution, including dealers, wholesalers, home centers, and builders. The company was formed in 1998, when several cabinet brands united as one. The oldest, Kemper, was founded in Richmond, Indiana, in 1926 and continues to be one of the company's high quality brands today. MasterBrand also sells cabinets under the HomeCrest, Kitchen Classics, Kitchen Craft, Maple Creek, Omega/Dynasty, Somersby, Orchard Park, and Value Choice brands.

MasterBrand's record of corporate and environmental stewardship is outstanding. The company is a certified manufacturer in the Kitchen Cabinet Manufacturers Association's Environmental Stewardship Program, which recognizes manufacturers that demonstrate an ongoing commitment to environmental practices and sustainability, and is equally committed to giving back to the communities where it operates by supporting academic and athletic youth programs, providing funds for Habitat for Humanity, and by donating to many other local and national causes.

MasterBrand Cabinets, Inc. attributes its success to great people and great brands, both of which are leading to a great future. The company will continue to provide superior quality cabinets at fair prices, honor every promise it makes, and help families turn their dreams into reality—because these values are at the heart of everything the company does and represent all that it stands for. Additional information about the company is available on its Web site at www.masterbrand.com.

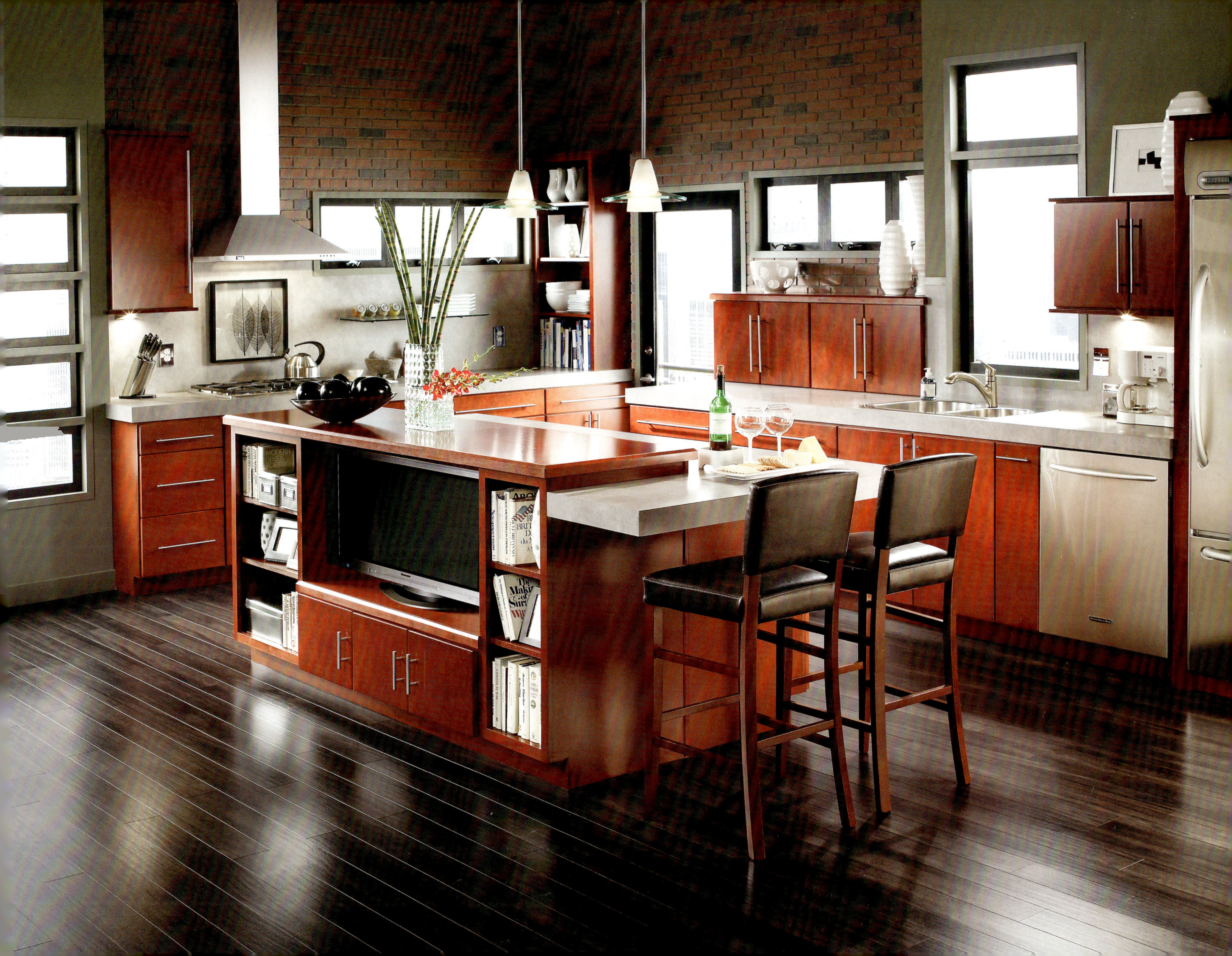

Kirby Risk Corporation

This multifaceted Lafayette-based company with more than 80 years of experience offers over 90,000 electrical and mechanical products and a variety of services and business processes, providing its clientele with one-stop shopping and responsive, customer-specific solutions designed to meet their needs.

Kirby Risk Corporation was founded in 1926 in Lafayette, Indiana, where it remains headquartered today. Kirby Risk responds to customer needs for solutions, quality service, and partnerships that help them achieve success. The company's products and services are delivered through its five business divisions: Kirby Risk Electrical Supply, Kirby Risk Service Center, Kirby Risk Precision Machining, Kirby Risk Mechanical Solutions and Service, and Arco Electric Products.

A Comprehensive Electrical Supply

Through Kirby Risk Electrical Supply, the company offers a full range of electrical products, including state-of-the-art automation control systems, providing its clientele with both quality and selection. With its 38 stocking locations, an overnight inventory transfer system, the use of electronic data interchange (EDI), and an online store with account access, Kirby Risk Electrical Supply also provides excellent customer service and technical support that is available 24 hours a day, seven days a week. Other services include inventory management and commercial and industrial lighting design.

A Capable Manufacturing Group

The Kirby Risk Service Center assembles and delivers wiring harnesses, subassemblies, custom-engineered systems, and just-in-time (JIT) kitting services. The service center builds and assembles to customers' exact engineering requirements and adds further value by helping to control component costs.

Kirby Risk Precision Machining responds to customers' needs with state-of-the-art computer numeric control (CNC) production technology. It provides quick turnaround and competitive pricing on precision-machined components for a wide range of industrial applications.

Arco Electric Products, located in Shelbyville, Indiana, designs and manufactures its own line of rotary phase converters and a broad offering of power factor correction equipment and bus bar manufacturing.

Serving mechanical needs, Kirby Risk Mechanical Solutions and Service offers a full range of power transmission equipment including customer support and technical assistance. It also performs repair and predictive maintenance services on both large and small electrical motors and apparatus to prevent equipment failures and expensive downtime.

Committed to meeting the electrical and mechanical needs of its customers, Kirby Risk continually seeks innovative and effective solutions. By supplying a wide range of products and providing a variety of services, Kirby Risk Corporation builds partnerships with its customers that lead to mutual success. Additional information is available on the company's Web site at www.kirbyrisk.com.

This page: Kirby Risk Corporation has branch locations throughout Indiana, Illinois, and Ohio. Opposite page: Kirby Risk is headquartered in Lafayette, Indiana.

KR Kirby Risk

Serving Indiana and beyond with responsive service... a Kirby Risk tradition since 1926

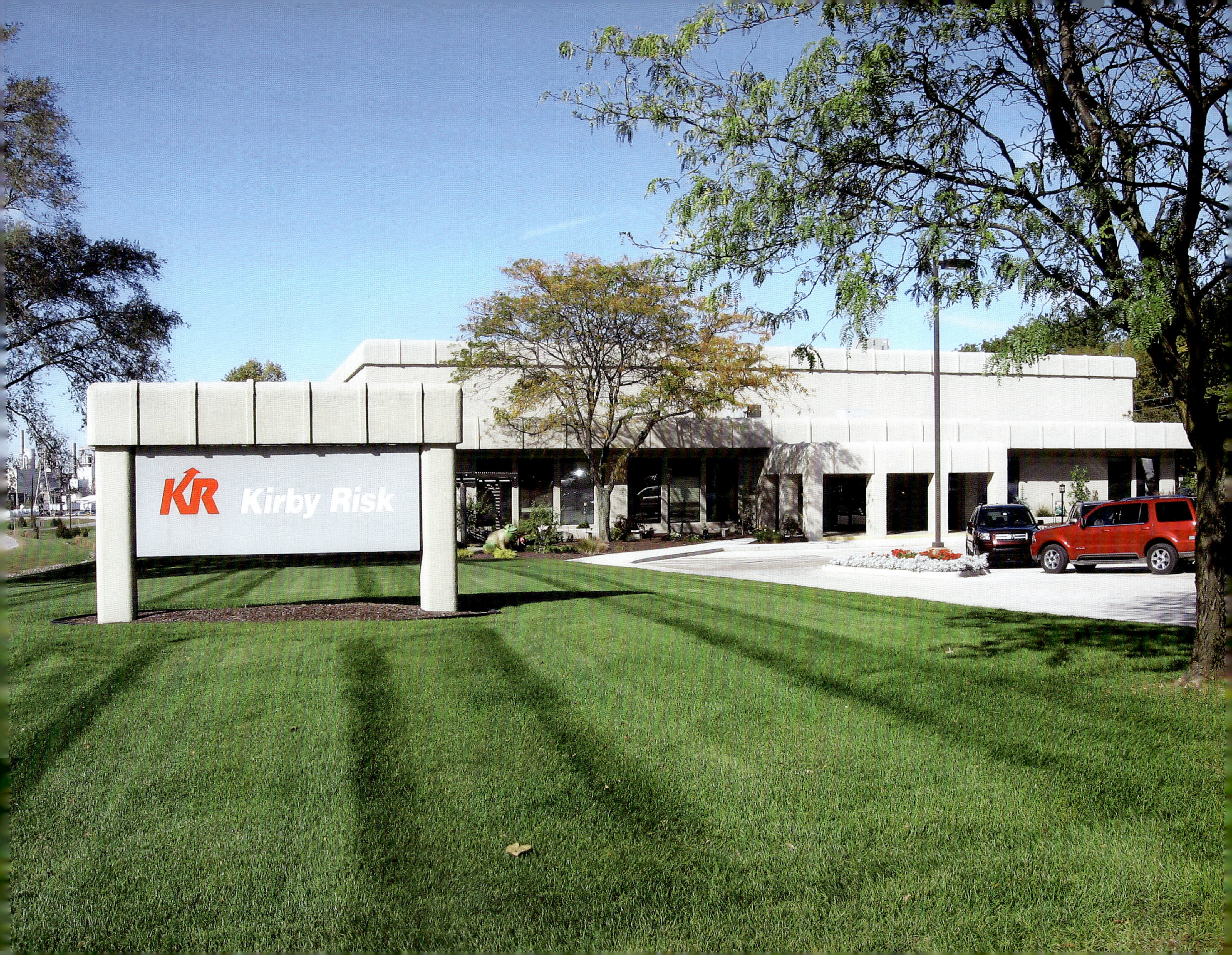
KR
Kirby Risk

Vertellus Specialties Inc.

This Indianapolis-based specialty chemical company is dedicated to manufacturing products that improve the quality of life. The company provides technical solutions for a variety of markets and is committed to the protection of the environment as well as to the safety and health of its employees and communities.

Above left: Vertellus Specialties Inc. is based in Indianapolis, Indiana. Above right: Vertellus chemists and engineers meet to discuss the latest project.

A leading provider of specialty chemicals, Vertellus Specialties Inc. uses innovative technology to develop market-focused solutions that create value for its customers and shareholders. Its products are used in the agriculture, nutrition, pharmaceutical, medical, personal care, plastics, coatings, and industrial markets.

Vertellus's business divisions include Agriculture and Nutrition Specialties, Health and Specialty Products, and Performance Materials. Agriculture and Nutrition Specialties is one of the market leaders in pyridine, used to increase agricultural productivity, and picolines, used in fungicides, herbicides, and insecticides as well as in rubber applications and the production of vitamin B3. Using a wide spectrum of complex technologies, the Health and Specialty Products division develops fine chemicals manufactured to meet customer needs. Performance Materials develops products derived from natural materials, such as castor, citric acid, and shea butter.

In July 2006, two leading chemical companies—Reilly Industries and Rutherford Chemicals—merged to form Vertellus Specialties Inc. These companies' histories include many important milestones in the chemical industry. For example, during World War II, when natural rubber supplies were decreasing, Reilly Industries' predecessor Reilly Tar & Chemical Corporation developed synthetic pyridine and key pyridine derivatives, which were used to manufacture synthetic rubber tires.

Today Vertellus employs about 750 people at commercial, manufacturing, and technology centers around the world to meet the demands of its customers. Corporate headquarters are located in Indianapolis, Indiana, as well as the manufacturing facility for pyridine, picolines, vitamin B3, corporate engineering, and research and development. The Indianapolis plant is one of the world's producers of pyridine and picolines, and like most of the company's sites, meets ISO-9001 standards.

Vertellus is dedicated to world-class health, safety, and environmental performance and participates in the global Responsible Care® initiative of the American Chemistry Council. Signatories to the initiative follow its guiding principles, which include supplying chemicals that can be safely manufactured, transported, used, and discarded; maintaining operations that protect the environment and contribute to the health and safety of employees and the public; and supporting education and research.

Vertellus Specialties Inc. will continue to pursue the responsible management of its chemical operations. This will be accomplished by using advanced technology in developing products and solutions that meet a variety of market needs and enhance quality of life. The company provides more information on its Web site at www.vertellus.com.

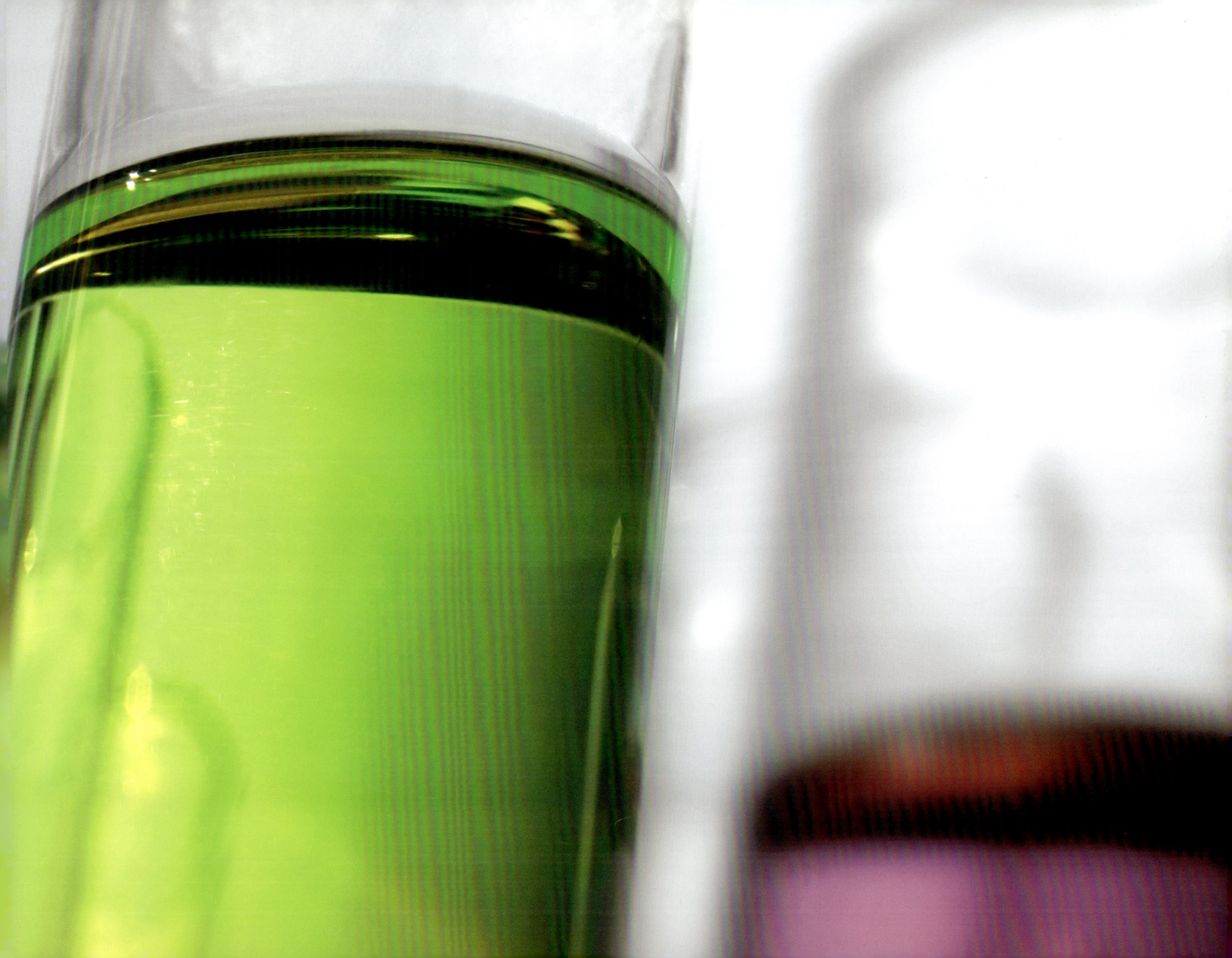

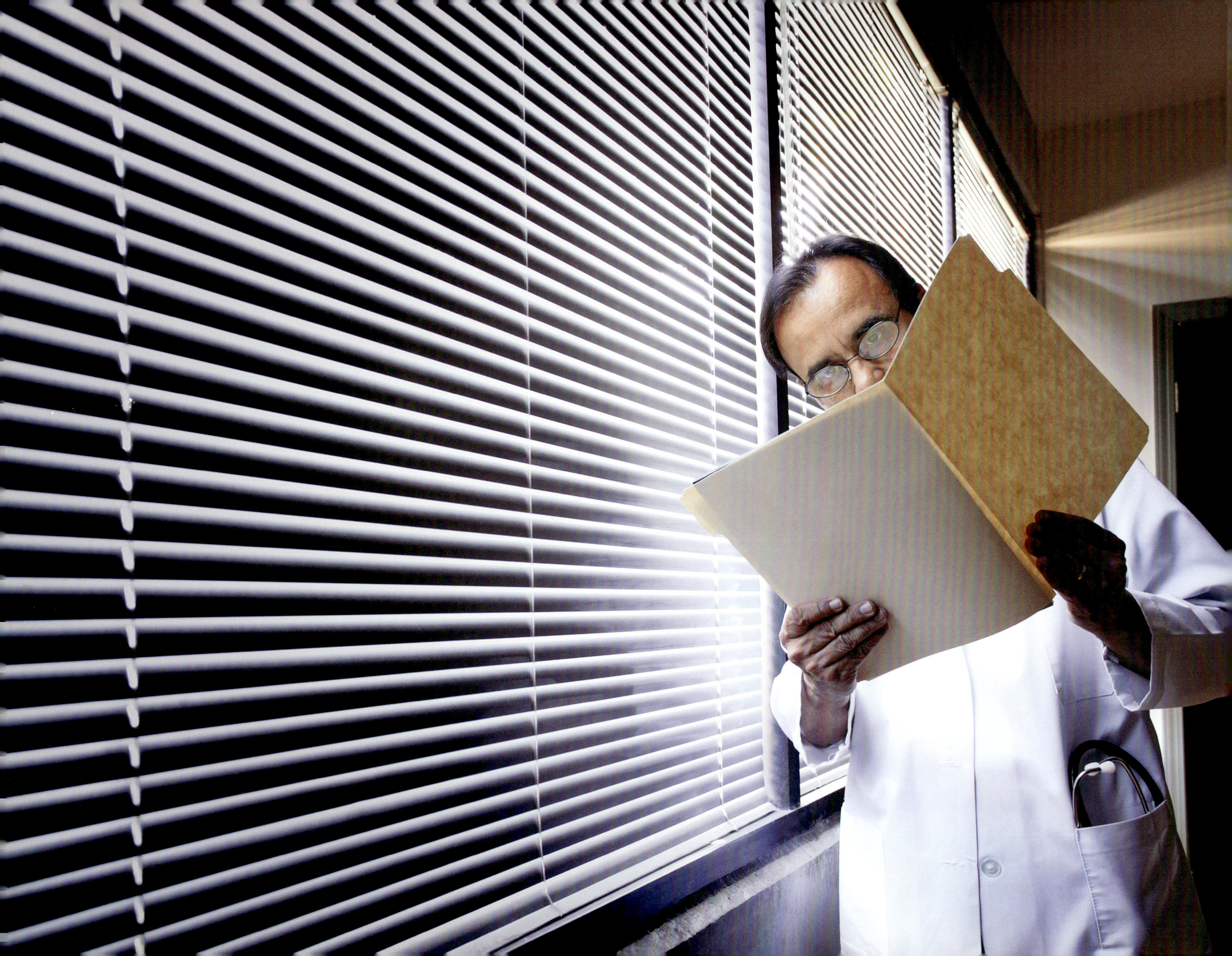

PROFILES OF COMPANIES AND ORGANIZATIONS
Pharmaceuticals, Medical Technology, and Health Care

Eli Lilly and Company

This research-based pharmaceuticals manufacturer, with one of the industry's strongest pipelines, is committed to supplying high quality medicines safely, with optimum environmental protection. Ranked as one of the best companies to work for, it values care for the community and sponsors programs to support patients worldwide.

Right: World headquarters for Eli Lilly and Company (Lilly) is in Indianapolis, Indiana. Flags surrounding the central drive represent some of the countries where Lilly maintains research, manufacturing, and sales presence.

Far right: Through murals, displays, and sculpture, the main lobby at Lilly headquarters reflects the company's passion for meeting urgent medical needs of people worldwide.

Eli Lilly and Company (Lilly), headquartered in Indianapolis, Indiana, is a leading innovation-driven biopharmaceutical corporation committed to developing a growing portfolio of best-in-class biopharmaceutical products that help people live longer, healthier, and more active lives. Lilly has been a publicly held company for more than 50 years. Lilly common stock is listed on the U.S. New York and Pacific Stock Exchanges (LLY), as well as on the London and Swiss Stock Exchanges.

Lilly's worldwide sales for 2007 were $18.63 billion. Lilly delivered outstanding results in 2007, highlighted by 14 percent pro forma sales growth, 17 percent growth in pro forma adjusted net income, and the movement into clinical testing of no fewer than 16 new molecules—unique compounds that are the foundation for potential medicines.

Lilly Products and Pipeline

The mission of Lilly is finding answers for some of the world's most urgent medical needs. Internal research efforts are focused primarily on four core therapeutic areas: neuroscience, endocrine disorders, cancer, and cardiovascular diseases. Lilly also continues to pursue innovative science and new opportunities beyond their targeted disease categories. Lilly pursues leading-edge science and technology from external and internal sources. Over the past decade, Lilly introduced important new drugs for the treatment of cancer, schizophrenia, osteoporosis, diabetes, cardiovascular complications, and severe sepsis. Lilly's mid-stage pipeline is the strongest in the company's history and has the potential to address many serious unmet medical needs of people around the world.

Lilly Philanthropy

Lilly also continues to build on its long tradition of philanthropy and community support. Lilly contributed more than $315 million in cash as well as products and other in-kind donations to charitable causes in 2007. This represents more than 6 percent of the Lilly adjusted income before taxes, ranking Lilly as one of the most generous companies in the world. Lilly also launched an employee volunteer initiative called Hands and Hearts, which will enhance employee involvement with nonprofit organizations. By encouraging volunteerism, Lilly strengthens its communities and increases the commitment and engagement of its workforce.

Lilly Manufacturing

As of 2007, Lilly global manufacturing operations were located in more than 20 sites in 12 countries and the U.S. territory of Puerto Rico. Lilly manufacturing consists of complex operations where employees translate deep knowledge of science, engineering, production, and numerous other disciplines into the reliable manufacture and supply of high quality, safe, and effective medicines. Because of the unique challenges associated with making medicine, Lilly places special emphasis on ensuring that its operations are safe in the broadest sense of the word—for patients, employees, the environment, and the communities in which Lilly operates.

Lilly Research and Development

The Lilly research division, Lilly Research Laboratories (LRL), conducts research in the United States, including Indiana and California, and more than 50 other countries. LRL is responsible for the discovery, development, clinical evaluation, and commercialization of potential new biopharmaceutical product therapies for many of the world's unmet medical needs.

During the research phase of the research and development process, Lilly targets diseases and sets priorities for the development of medicines. Lilly seeks to develop biopharmaceuticals that are "first in class" and/or "best in class." Lilly is continuing to pursue its corporate strategy of tailored therapeutics to provide greater value and predictability from biopharmaceuticals. Tailored therapeutics use a variety of emerging tools and technologies to identify the patients most likely to benefit from a medicine and to tailor the dose, timing, and other factors to achieve better outcomes. Lilly has been reorienting its research and development process to further tailor its products in order to reduce the time and investment required to bring products to market.

Elanco Animal Health

A division of Lilly, Elanco Animal Health markets products worldwide to improve the health of animals. Products are marketed primarily to cattle, poultry, and swine producers. Elanco has key operations in North America, Europe, the Middle East, Africa, the Asia-Pacific area, and Latin America and makes its products in more than 100 countries. In 2007 Elanco entered the companion animal health business with product plans for cats and dogs.

Elanco's product line concentrates on four areas: antibacterials, parasiticides, anticoccidials, and productivity enhancers. The division develops and markets innovative technologies for use in animal production, care, disease treatment, and prevention.

More Information

The company describes its corporate history and its significant medical breakthroughs on the About Us page on the Eli Lilly and Company Web site at www.lilly.com.

Above left: The Lilly research division—Lilly Research Laboratories (LRL)—conducts pharmaceutical research and development in the United States, including Indiana and California, and more than 50 other countries. Above right: An LRL scientist conducts lead optimization, the early stage of pharmaceutical research during which leads (molecules that are the foundation for potential medicines) are processed to optimize their biochemical characteristics.

Roche Diagnostics

This Indianapolis-based company is a world leader in the diagnostics industry and is committed to improving patient care by advancing diagnostics tools. It supplies its broad portfolio of innovative products and services to laboratories, physicians, and patients around the world.

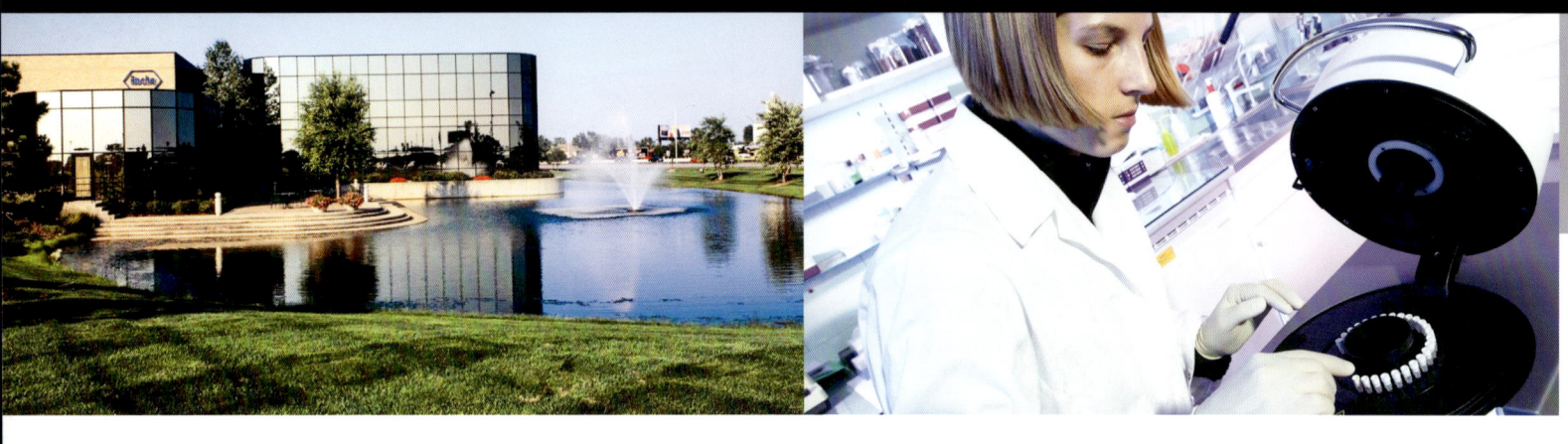

Above, left to right: Roche Diagnostics offers a scenic campus for its employees. Roche's LightCycler® Carousel-Based System was the first system to introduce hybridization probes. The ACCU-CHEK® Aviva blood glucose meter was launched in 2005. Roche holds several patents for the Osteoc E170, which is used in osteoporosis testing. The AmpliChip CYP450 Test analyzes variations in two genes that play a major role in the metabolism of many widely prescribed drugs.

Roche Diagnostics is dedicated to finding and developing new solutions to diagnose, treat, and prevent disease. In 2006 Roche Diagnostics demonstrated its dedication to innovation when it invested 9 percent—$557 million—of its global sales in research and development. It continually strives to improve its numerous products, which deliver actionable health information—data that includes a patient's predisposition to different diseases and the effectiveness of certain therapies—that allows both physicians and patients to make informed health care decisions. The company's expertise, innovative thinking, and global presence help drive the diagnostics industry.

In 1998 Roche Diagnostics acquired Boehringer Mannheim Corporation (BMC), based in Indianapolis and a leader in the diagnostics industry. With this acquisition, Roche Diagnostics became the world's largest in-vitro diagnostics company and established its North American headquarters in Indianapolis. It currently employs more than 3,700 people at this location.

In 2005 Roche Diagnostics' parent company, the Roche Group, celebrated its 100th anniversary of health care innovation in the United States. Roche Diagnostics' large array of products includes many innovations, such as the ACCU-CHEK product line, Polymerase Chain Reaction (PCR) technology, and the AmpliChip CYP450 Test.

The ACCU-CHEK meter—developed by BMC—was introduced in 1982 as the first device that allowed patients to monitor their own blood glucose levels. Roche Diagnostics continues to expand this extensive product line, helping people to manage their diabetes. In 1991 Roche acquired the worldwide rights and patents for PCR technology, which can multiply genetic material rapidly. This Nobel Prize–winning advancement has led to many other significant developments in genetic studies. Roche currently holds more than 130 patents related to PCR. In 2005 the FDA approved the AmpliChip CYP450 Test —a world first in the medical industry— which helps predict the rate at which a patient metabolizes certain drugs. This test can help increase drug efficacy rates and reduce the number of severe drug reactions.

Roche Diagnostics focuses on people, performance, and leadership. This focus is reflected in the company's inclusion in *Fortune*'s list of the 100 best companies to work for in 2005. Like its parent company, Roche Diagnostics believes that exemplary business practice includes social responsibility. The Roche Group was one of the first U.S. companies to begin a corporate giving program. Caring for people drives Roche Diagnostics to continue to develop advanced products that help improve patient care and quality of life. Roche Diagnostics provides additional information on its Web site at www.roche-diagnostics.us.

ACCU-CHEK, AMPLICHIP, and LIGHTCYCLER are trademarks of Roche. The technology used for the LightCycler is licensed from Idaho Technology, Inc.

Roche
AMPLICHIP
CYP450 ARRAY
Powered by Affymetrix
For Research Use Only
Not for use in diagnostic procedures

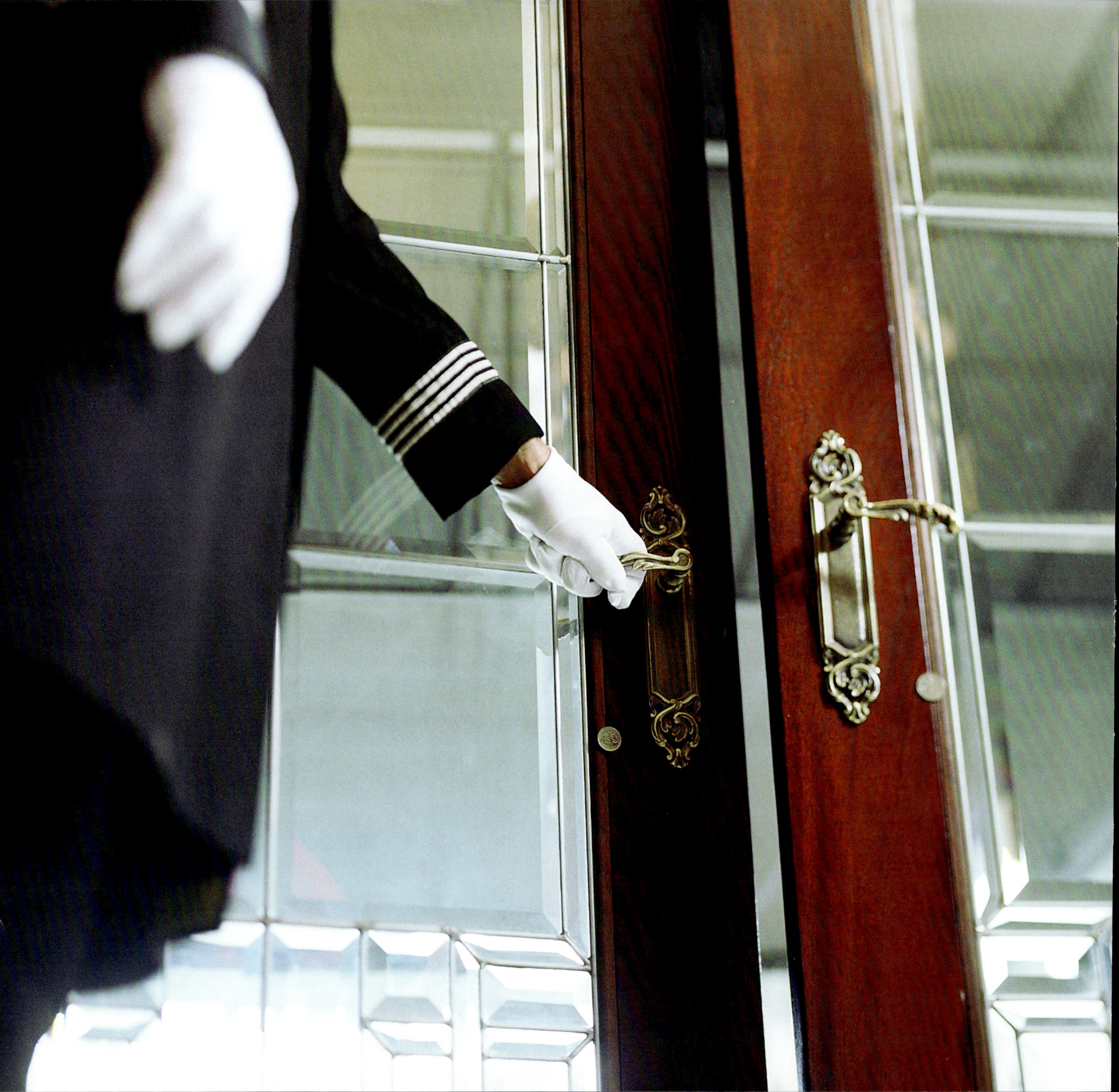

PROFILES OF COMPANIES AND ORGANIZATIONS
Tourism, Entertainment, and Hospitality

French Lick Resort

A traveler's paradise, this artfully restored resort with natural mineral springs was a 1920s haven for the celebrities and political figures of the era. Today it is an elegantly appointed destination offering every modern guest amenity, luxurious spas, and diverse recreational venues, and contributing generously to the community.

Above: The luxurious French Lick Springs Hotel is a historic highlight of the French Lick Resort, which also includes West Baden Springs Hotel and French Lick Casino.

Long known for its healing natural mineral springs and signature-designed golf courses, the French Lick Resort today is the culmination of a $500 million restoration and casino development project. The premier resort includes West Baden Springs Hotel, French Lick Springs Hotel, and French Lick Casino. The destination features 689 guest rooms and suites, a 51,000-square-foot casino, a 60,000-square-foot sports center, 45-holes of championship golf, two full-service spas, 14 restaurants, horseback riding stables and trails, a 109,000-square-foot Conference and Event Center, and a fascinating past that reads like a who's who of the rich and famous.

French Lick Springs Hotel

Built in 1901, the French Lick Springs Hotel is on the National Register of Historic Places. More than 440 guest rooms have been renovated to offer the luxury and modern amenities of today while also reflecting the old-time charm of the hotel's glorious past. The Conference and Event Center features a grand ballroom and junior ballroom with outdoor terraces that overlook the hotel's formal gardens.

The Spa at French Lick offers curative and aesthetic treatments, while a fitness center and indoor/outdoor swimming pools and hot tub complex invite more active endeavors. Added to the list of diverse venues are three championship-level golf courses, the hotel's state-of-the-art sports center, a bowling alley, several retail shops, and the casino. In addition, guests are offered a choice of nine restaurants.

The beaux arts–style French Lick Casino is directly connected to the French Lick Springs Hotel by an enclosed, climate-controlled walkway. The gaming floor offers 1,345 slot machines and 41 gaming tables, including blackjack, roulette, craps, and 10-seat electronic Texas Hold 'Em poker tables, as well as a high-limit VIP gaming area. Dining choices include an entertainment lounge and sports bar, a bistro, and a casual delicatessen-style eatery.

West Baden Springs Hotel

The West Baden Springs Hotel, a National Historic Landmark (NHL), is a crown jewel of the French Lick Resort. Built as a health resort in 1902, the hotel underwent a major restoration and in 2007 welcomed guests into newly refurbished grand hotel rooms and suites. Renovated to reflect an era of glamorous celebrities, roaring jazz, and opulent luxury, the hotel features a mix of elegant furnishings and fabrics from around the world. The Spa at West Baden is a two-story natatorium recreated from the original 1902 design and includes spa treatments, pools, and a fitness center.

A spectacular atrium at West Baden Springs Hotel was once the world's largest free-span dome and was dubbed "The Eighth Wonder of the World." West Baden Springs Hotel offers an ideal setting for board meetings, executive retreats, and other gatherings.

A Golfer's Paradise

In a perfect world, any golf enthusiast would be able to play on courses designed by Donald Ross and Pete Dye at one exceptional resort. French Lick offers just that sort of golfing heaven. One of the Midwest's premier golf destinations, the resort's restored 1917 Donald Ross Course offers a traditional playing experience augmented with modern technologies. The newly designed Pete Dye Course is a championship-level, 18-hole golfing masterpiece that blends into the rolling bluffs and hillsides behind the French Lick Springs Hotel. The nine-hole Valley Links gives golfers of all skill levels an enjoyable and challenging place to play and is located within walking distance of the hotel's sports center, with indoor putting green, driving ranges, short game practice greens, and golf lessons.

The Spas' Healing Serenity

True to its past role as one of the nation's grand health resorts, The Spas at French Lick and West Baden provide guests with relaxing surroundings and the healing minerals of the local spring waters. Services at both facilities offer mineral baths, hydrotherapy treatments, massage, manicures, pedicures, facials, and haircuts, and French Lick also includes herbal wraps. Italian limestone, stained-glass windows, and wood furnishings bring a touch of old-world charm to the refurbished spas.

The resort provides additional information about all of its facilities and services on its Web site (www.frenchlick.com).

Valuable Preservation

At $500 million, the resort's renovation became the largest privately funded restoration project in Indiana history and one of the largest projects of its kind in the United States. The hotel's historic grandeur has been retained while its amenities offer the level of service and luxury expected by guests in the 21st century.

As the French Lick Resort restoration has preserved a valuable hotel and recreational facility, the resort's transformation has increased employment in the community of French Lick. The resort has helped to improve the quality of life in Orange County by providing jobs, promoting business, and making monetary donations to county schools.

Casino revenues to the town of French Lick have also enabled the purchase of police vehicles, the improvement and beautification of community areas, and the start of a downtown development project slated to include dining, retail, residential housing, and a town green. Whether by donating to local charitable causes or bringing jobs to Indiana residents, the French Lick Resort has a continuing positive impact on the community.

Above right: West Baden Springs Hotel, built in 1902 as a sophisticated health resort, has now been elegantly refurbished. Above left: The hotel's spacious atrium is topped by a 200-foot, free-span dome that was at one time the largest in the world.

Hyatt Place Indianapolis Airport

Newly designed for today's relaxed lifestyle, this distinctive hotel—conveniently located near the Indianapolis International Airport—offers busy travelers a hospitality Gallery and café, flat-panel HDTVs, advanced fitness equipment, high-speed Internet access, and comfortable guest rooms with every contemporary amenity.

Above left: The Hyatt Place Indianapolis Airport's Gallery has a contemporary design and includes a den with a flat-panel television; the Gallery Café with premium coffees, beer, and wine and an array of snacks and gourmet baked items; and an e-room that offers free computers and printers. Top right: Hotel guests can order freshly prepared food 24 hours a day, seven days a week via a touch screen menu in the Guest Kitchen.

Located at the Crossroads of America, the Hyatt Place Indianapolis Airport is a new kind of hotel that puts style, innovation, and The Hyatt Touch within everyone's reach. Just minutes from the Indianapolis International Airport, the Hyatt Place offers complimentary 24-hour airport shuttle service and is conveniently located near some of the best shopping and entertainment in the area. It provides easy access to downtown Indianapolis and popular attractions such as the Indianapolis Motor Speedway, the Indiana Convention Center and Lucas Oil Stadium, the Indianapolis Museum of Art, the Children's Museum of Indianapolis, Metropolis mall, Circle Centre mall, the historic Indianapolis City Market, the Indianapolis Zoo and White River Gardens, and White River State Park.

Spacious Hyatt Place Indianapolis Airport guest rooms offer special amenities to make traveling a pleasure. Among these are a sitting area with a Cozy Corner oversized sofa-sleeper; a 42-inch flat-panel, high-definition television (HDTV) with movies, sports, games, and music on demand; a Hyatt Plug Panel for easily connecting media devices such as laptop computers, MP3 players, video game systems, and DVD players directly to the TV; a wet bar; a plush Hyatt Grand Bed with thick pillow-top mattress and down pillows; and spa products for the bath. There is also a flexible workspace with task lighting, plus coffeemaker and refrigerator.

The Hyatt Place Indianapolis Airport offers a common area that is as warm and inviting as the hotel's guest rooms. Called the "Gallery," this is a casual environment where guests may take advantage of the express check-in/check-out kiosk; work, relax, and socialize in comfortable surroundings; enjoy coffee, wine and beer, juices, and energy drinks in the Gallery Café; or be entertained in the television den. The e-room provides a complimentary computer and printer. Among other amenities are a StayFit@Hyatt fitness center with Life Fitness cardiovascular equipment that includes television consoles; an outdoor swimming pool; complimentary high-speed Wi-Fi Internet access throughout the hotel; and Gallery Hosts to assist guests around the clock.

The Hyatt Place Indianapolis Airport offers multiple dining options. To begin the day, there is a complimentary continental breakfast, which features fresh fruit and juice, cereal, toast, bagels, yogurt, and coffee, or guests may purchase made-to-order hot, American-style breakfasts and Starbucks specialty coffees. Meals and snacks are also available 24 hours a day, seven days a week, including freshly prepared entrees, sandwiches, soups, salads, and pizza, all of which may be ordered through the Gallery Host or via a cashless kiosk with touch-screen menu in the Guest Kitchen.

For corporate meetings and private events, the Hyatt Place Indianapolis Airport offers an all-inclusive meeting package, with state-of-the-art audiovisual equipment, and fine food and beverages. A Meeting Host who is committed to the success of the event will assist with planning and be on hand to ensure that the premises are set up to meet the client's needs, including conference tables and chairs, wired and wireless projectors and a projection screen, a 42-inch flat-panel HDTV, speakerphone, easels, and more. More information is available on the hotel's Web site (indianapolisairport.place.hyatt.com).

Hyatt Place Indianapolis Keystone

This stylish hotel offers contemporary accoutrements for on-the-go travelers, with state-of-the-art meeting facilities, free Wi-Fi Internet access, guest rooms with full amenities, satellite HDTV, round-the-clock fresh food options, specialty coffees, and superb service—all close to shopping, dining, museums, sports, and more.

At the Hyatt Place Indianapolis Keystone—located in the fashionable Keystone at the Crossing area, 20 minutes from downtown Indianapolis in the city's far northside neighborhood—guests are treated to a long list of amenities that make business trips and vacations something to be excited about. The hotel's 2008 renovation is a thoroughly modern success that incorporates The Hyatt Touch into every corner of the hotel and all areas of guest service.

At the Hyatt Place Indianapolis Keystone, no detail is overlooked. For meetings, events, and private parties, the hotel provides an all-inclusive package. Events and meetings are set up to conform to the client's needs through a Meeting Host who handles all the details, including state-of-the-art audiovisual equipment, conference tables and chairs, wired and wireless projectors and a projector screen, a 42-inch flat-panel, high-definition television (HDTV), and other amenities.

Beyond the conference room, first-class hospitality extends into the heart of the hotel—the Gallery. The Gallery combines the convenience of a lobby with the warm, elegant atmosphere of a Great Room. The spacious setting provides a welcoming area where guests can relax or work or chat with friends. It also features the Gallery Café, serving coffee, wine and beer, juices, and energy drinks; a television viewing area with large, flat-panel HDTV; and an e-room with a computer and printer for guests to use. Wireless and wired high-speed Internet access are complimentary throughout the hotel.

Gallery Hosts are available around the clock to assist guests with check-in/check-out and travel details and provide information about local activities and tours. They can help with tickets, tee times, reservations, and recommendations. The Hyatt Place Indianapolis Keystone is close to myriad attractions for entertainment and enjoyment, including the Indianapolis Motor Speedway and other sports venues, museums and a zoo, shopping malls, and nature preserves. The hotel is also close to many fine restaurants, and on the grounds of the hotel there is a state-of-the-art fitness center and a relaxing outdoor swimming pool and lounge area.

Each guest room at the hotel incorporates the features of smart, contemporary living and home-style comfort. The signature Grand Hyatt Bed, with luxuriously thick mattress and down pillows, is a pleasure to sink in to. In the sitting area, which is separated from the main sleeping area, is a queen-size sofa-sleeper and well-lighted desk area. Wireless and wired high-speed Internet access are complimentary throughout the hotel. A 42-inch, flat-panel HDTV is compatible with laptop computers, MP3 players, video game systems, and DVD players. Satellite channels include SportsNet and other professional and college sports options, movies, games, and music on demand.

Dining options at the Hyatt Place Indianapolis Keystone have also been revamped to offer the foods and beverages guests want wrapped in great taste and convenience. Visitors may enjoy a complimentary continental breakfast or purchase an American-style breakfast and Starbucks coffee. In addition, made-to-order soups, sandwiches, salads, pizza, and entrees are available for purchase 24 hours a day from the Guest Kitchen. Additional information is available on the hotel's Web site at www.indianapoliskeystone.place.hyatt.com.

Top right: The Hyatt Place Indianapolis Keystone's spacious king guest room features stylish furnishings, a state-of-the-art media and work center, a high-definition television, a sofa-sleeper, and a wet bar. Above: The hotel's Gallery is a warm and open area that features a self-registration kiosk, an intimate coffee and wine café, a television den, and an e-room with free access to a public computer and printer.

The Courtyard by Marriott Indianapolis–Carmel

Notably praised by its guests and the hospitality industry, this hotel is located just 20 minutes from downtown Indianapolis and 10 minutes from downtown Carmel. It is close to the historic Old Northside area, to shopping and dining venues, and to a variety of attractions, making it an ideal choice for business and leisure travelers.

Above left: The Courtyard by Marriott Indianapolis–Carmel provides a spacious setting and a convenient location for guests traveling for business or pleasure.

Top right: Guests are offered a welcoming, smoke-free environment throughout the hotel. The warm, comfortable lobby conversation area invites relaxation or visits with friends, family members, or business clientele.

Bottom right: Guest rooms and suites are newly renovated and are appointed with fine bedding, including custom-designed comforters, thick mattresses, and fluffy pillows.

The Courtyard by Marriott Indianapolis–Carmel, located in Indianapolis, Indiana, is a newly renovated hotel offering 149 rooms, including 12 suites. The $2.5 million project, completed in October 2006, revamped all of the hotel's public areas and guest rooms, which are appointed with new furniture, and for the guest rooms, new luxury bedding, oversized mattresses, and luxurious pillows. The spacious guest rooms and suites feature fully remodeled, modern bathrooms; ergonomically designed, well-lighted work spaces; two telephones with multiple lines, data ports, and voice mail to accommodate business transactions. Complimentary high-speed wireless Internet access is offered in guest rooms and throughout the hotel, along with the standard complimentary high-speed wired access.

Amenities and Attractions

The entire hotel provides a smoke-free environment for the comfort of its guests. On-site is the Courtyard Café, a full-service, breakfast-only restaurant featuring a generous buffet as well as cooked-to-order entrees. For guests' convenience, there is the Market, a self-serve shop open around the clock, offering food, snacks, and beverages. There are also numerous restaurants nearby.

Additional amenities at the hotel include a guest lounge where guests can gather to relax, enjoy a large-screen television, converse, or dine. For business travelers, the hotel has created the Business Library Center, a full-service business area that features individual workstations and ergonomic chairs. The hotel also provides two meeting rooms, each accommodating up to 50 people. After a busy working day, guests can rejuvenate in the hotel's indoor swimming pool and whirlpool spa or at the fitness center, which includes weight-training equipment. Furthermore, guests have complimentary use of Lifestyle Family Fitness, a full-service gym located within one block of the hotel. And of course, the Marriott Rewards program makes travel a little more rewarding.

For recreation and sightseeing, the hotel is within five miles of four golf courses and partners with a number of local courses to offer "Stay and Play" packages. Many local attractions are also in close proximity, including the Carmel Arts and Design District; Central Park and Aquatic Center; Monon Center and Trail; Clay Terrace and Keystone at the Crossing shopping, dining, and entertainment centers; Conner Prairie Living History Museum; Indianapolis Museum of Art; the Children's Museum of Indianapolis; the Indianapolis Zoo; White River State Park and its attractions; and the Indianapolis Motor Speedway and Hall of Fame Museum.

The Courtyard by Marriott Indianapolis–Carmel provides additional information about the hotel's features on its Web sites (www.indycarmelcourtyard.com and www.marriott.com).

Awards and Accolades

The Courtyard by Marriott Indianapolis–Carmel has won numerous notable honors from Marriott International, including a 2007 Gold Hotel Award in guest satisfaction; "hotel of the quarter" in Marriott International's Central Region for the second and third quarters of 2007; and the highest overall guest satisfaction rating for the Marriott Courtyard brand in the Central Region in 2007.

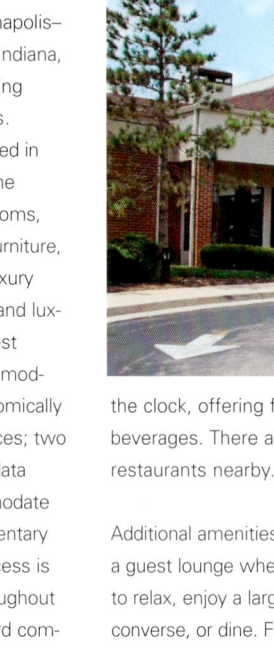

All photos: © Ben Bremen

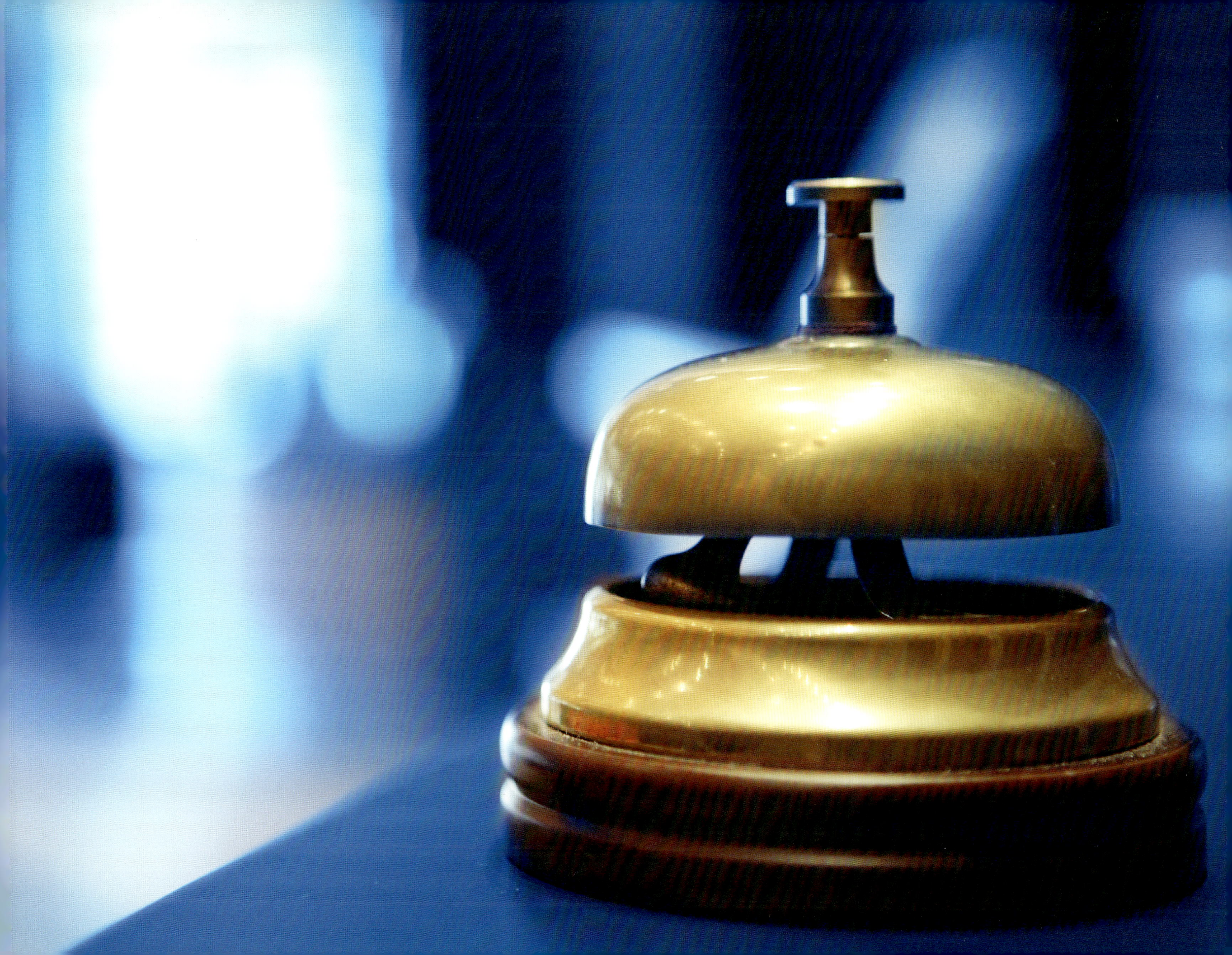

PHOTO CREDITS

Page ii: © Geoff Hill
Page v: © Jason Lindsey/Alamy
Page vi: © Charles Farris
Page viii, left: © Carl Van Rooy
Page viii, ix: © Richard Cummins
Page ix, right: © Matthew C. Hale
Page x: © Ron Hoskins/NBAE via Getty Images
Page x, xi: © Marcy Kellar
Page xi, right: © Andy Attenburger/Icon SMI/Corbis
Page xii, left: © Joseph Marinaro
Page xii, xiii: © Jon Brewer
Page xiii, right: © Gary Brockman
Page xiv, left: © Adam Schweigert
Page xiv, xv: © Adam Schweigert
Page xv, right: © Ken Horn
Page xvi, left: © Jeff Greenberg/age fotostock
Page xvi, xvii: © Kevin Foster
Page xviii, left: © Dennis MacDonald/Alamy
Page xviii, xix: © Bruce Foster/Getty Images
Page xix, right: Courtesy, The Indiana Rail Road Company. Photo by Mike Stickel
Page xxi: © Getty Images
Page xxii: Courtesy, HOK. Photo by Sam Fentress
Page 2: © Samantha McGranahan
Page 4: © Thomas Damgaard Sabo
Page 5: © Bart Roberts
Page 6, left: © Daniel Dempster Photography/Alamy
Page 6, right: © Aldo Murillo
Page 7: © William Kerney
Page 8, top: © Gary Brockman
Page 8, bottom: © Brian Thome
Page 9: © Pendleton-Gazette Video News Magazine
Page 10: Courtesy, Indiana Office of Tourism Development
Page 11: © Pavel Trebukov
Page 12: © Shane Pequignot/Eye Pix Photography
Page 13, left: © Warren Lynn
Page 13, right: © Andy Altenburger/Icon SMI/Corbis

Page 14, left: © WayNet.org
Page 14, right: © Sherry L. VerWey
Page 15: © George Tiedemann/GT Images/Corbis
Page 16: © Thomas Gill
Page 17, left: © Brian Kelly
Page 17, right: © Andrew Swanson
Page 18, left: © Yo Hibino
Page 18, right: Courtesy, Indiana University
Page 19: © Kord.com/age fotostock
Page 20: © Eric Murray
Page 21, left: © Ryan Caiazzo
Page 21, right: Courtesy, French Lick Resort & Casino
Page 22: © Phil Cardamone
Page 23: Courtesy, Toyota Motor North America, Inc.
Page 24: Courtesy, Purdue University. Photo by David Umberger
Page 25, left: Courtesy, General Motors
Page 25, right: Courtesy, Jayco
Page 26, left: Courtesy, Navistar
Page 26, 27: Courtesy, Delphi Corp
Page 27, right: Courtesy, Chrysler LLC
Page 28, left: Courtesy, United States Steel Corporation
Page 28, right: Courtesy, Steel Dynamics, Inc.
Page 29: Courtesy, SABIC Innovative Plastics
Page 30, left: Courtesy, Jofco Inc.
Page 30, right: Courtesy, MasterBrand Cabinets Inc.
Page 31, left: Courtesy, Old Hickory Furniture Company
Page 31, right: © Lee Davenport
Page 32: © Chris Schmidt
Page 33, left: © AP Photo/Jack Dempsey
Page 33, right: Courtesy, Sony DADC
Page 34: © Scott Sinklier/Corbis
Page 35: © Marcelo Wain
Page 36, left: © Macduff Everton/Corbis
Page 36, right: Courtesy, Young & Laramore/Dale Bernstein Photography
Page 37: Courtesy, U.S. Department of Defense. Photo by Petty Officer 2nd Class Maebel Tinoko, USN
Page 38, left: © Nenad Colovic

Page 38, right: © AP Photo/Larry Crowe
Page 39: © Stephen L. Parker/aroundfortwayne.com
Page 40: Courtesy, Rose Acre Farms. Photo by John Rust
Page 41, left: © AP Photo/The Indianapolis Star/Adriane Jaeckle
Page 41, right: Courtesy, Indiana Packers Corporation, Delphi Indiana
Page 42, left: © AP Photo/Ron Kuenstler
Page 42, right: © AP Photo/Lisa Poole
Page 43: Courtesy, Dow AgroSciences LLC
Page 44: © Andy Altenburger/Icon SMI/Corbis
Page 45: Courtesy, Purdue University
Page 46, left: © Gary Fuss
Page 46, right: Courtesy, Indiana University–Purdue University Indianapolis
Page 47: Courtesy, Indiana University–Purdue University Indianapolis
Page 48, left: Courtesy, Purdue University. Photo by David Umberger
Page 48, 49: Courtesy, Vincennes University
Page 49, right: © Britton Starr
Page 50, left: Courtesy, Indiana State University. Photo by Drew Lurker
Page 50, 51: © AP Photo/Joe Raymond
Page 51, right: Courtesy, Hanover College
Page 52, left: Courtesy, Jim Amidon/Wabash College
Page 52, right: Courtesy, Marshall Space Flight Center/NASA
Page 53: Courtesy, Ivy Tech Community College
Page 54: © Andrew Douglas/Masterfile
Page 55: © AP Photo/Darron Cummings
Page 56, left: Courtesy, Eli Lilly and Company
Page 56, 57: © AP Photo/Mark Lennihan
Page 57, right: © Jody Dole/Getty Images
Page 58, left: Courtesy, Cook Pharmica LLC
Page 58, right: Courtesy, Roche Diagnostics
Page 59, left: © AP Photo/Darron Cummings
Page 59, right: Courtesy, Biomet, Inc
Page 60, left: Courtesy, Purdue University. Photo by David Umberger
Page 60, right: Courtesy, Purdue University
Page 61: Courtesy, Indiana University. Photo by Jacob Kriese
Page 62: Courtesy, LifeLine, Clarian Clinical Care Transport Service
Page 63, left: © AP Photo/Darron Cummings
Page 63, right: © AP Photo/AJ Mast

Page 64: Courtesy, Celadon Group Inc.
Page 65: © Daniel Vincent
Page 66: © AP Photo/Michael Conroy
Page 67: © Jordan Barclay
Page 68: Courtesy, Indianapolis Airport Authority
Page 69: © Stephen Dobbins
Page 70: © David R. Frazier Photolibrary, Inc./Alamy
Page 71: © David Mendelsohn/Masterfile
Page 72: © AP Photo/M. Spencer Green
Page 73, left: Courtesy, NiSource Inc.
Page 73, right: © Cindy Seigle
Page 74: © Pixelhouse/Corbis
Page 75: © Super Stock
Page 76, left: Courtesy, Simon Property Group
Page 76, right: Courtesy, Simon Property Group
Page 77: Courtesy, Duke Realty Corporation
Page 78: © Murphey Studios Inc.
Page 79, left: Courtesy, Hunt Construction
Page 79, right: © Paul D. Van Hoy II
Page 80: © Paul D. Van Hoy II
Page 81: © Thiery Dosogne/Getty Images
Page 82: © Michael Downey
Page 83, left: © Catherine Yeulet
Page 83, right: © Shih-Pei Chang
Page 84: © Lajos Répási
Page 86: © Marc Dietrich
Page 90: © Tim Pannell/Corbis
Page 94: © Timmy Photo
Page 101: © JR Donnolly
Page 103: © TexPhoto
Page 104: © Andreea Manciu
Page 108: © Alvaro Heinzein
Page 120: © Sabrina dei Nobili
Page 122: © MB Photo
Page 127: Courtesy, Roche Diagnostics
Page 128: © Todd Pearson/Getty images
Page 135: © Eva Serrabassa

cherbo publishing group, inc.

TYPOGRAPHY
Principal faces used: Univers, designed by Adrian Frutiger in 1957;
Helvetica, designed by Matthew Carter, Edouard Hoffmann,
and Max Miedinger in 1959

HARDWARE
Macintosh G5 desktops, digital color laser printing with Xerox Docucolor 250, digital imaging with Creo EverSmart Supreme

SOFTWARE
QuarkXPress, Adobe Illustrator, Adobe Photoshop, Adobe Acrobat, Microsoft Word, Eye-One Pro by Gretagmacbeth, Creo Oxygen, FlightCheck

PAPER
Text Paper: #80 Luna Matte

Bound in Rainbow® recycled content papers from Ecological Fibers, Inc.

Dust Jacket: #100 Sterling-Litho Gloss